★ ★ A HISTORY OF ★ ★
SUBMARINE WARFARE
══ ALONG ══
THE JERSEY SHORE

Joseph G. Bilby & Harry Ziegler

H
THE
History
PRESS

Published by The History Press
Charleston, SC
www.historypress.net

First published 2016

Manufactured in the United States

ISBN 978.1.46713.526.9

Library of Congress Control Number: 2016932971

Dedicated to the memory of Patricia Ann Ziegler Bilby, 1947–2015, beloved wife and sister.

CONTENTS

ACKNOWLEDGEMENTS

No book is created by its authors alone, and we would like to thank those people who made this work possible and provided information and research leads that made for a much better book. They include John Allocca, Jay Amberg, Allen Atheras, Margaret Thomas Buchholz, Steve Garratano, Cory James Newman, Eric Olsen, Damian Shiels and Marjy Wienkop. If there is anyone whom we have forgotten, we extend our apologies.

INTRODUCTION

New Jersey has a long history with submarines. The first American submarine, David Bushnell's *Turtle*, which failed in its attempt to sink a British vessel off Manhattan, met its end in 1776 when the ship on which it was being transported to New Jersey went to the bottom of the Hudson River off Fort Lee. The *Intelligent Whale*, the only surviving Union Civil War submarine, which was sponsored by New Jersey investors, largely built in Newark and owned by New Jerseyan Oliver Halsted, sits in the National Guard Militia Museum of New Jersey in Sea Girt.

The first successful submarine adopted by the United States Navy was invented by Paterson Irish immigrant John Holland, who tested his prototype undersea boats in the Passaic River and in the Hudson River on the Jersey City waterfront. His final model was built in Elizabeth.

The state's largely benign relationship with submarines turned sour in 1918, however, when the German navy's *U-151* went on a one-day six-ship sinking rampage off the New Jersey coast. Worse was yet to come. A World War II U-boat offensive, nicknamed by German submariners as "The Happy Time," torpedoed numerous ships off New Jersey in 1942, leaving oil-soaked beaches strewn with wreckage and an occasional body as the state government and the military struggled with a response—and each other. That coastal conflict has left an echoing narrative that resonates dimly down to the present day. In these pages we tell that little-known and long-forgotten story.

Chapter 1

NEW JERSEY AND EARLY SUBMARINES

"THE PROJECT WAS PRACTICALLY IMPOSSIBLE"

New Jersey is actually a peninsula, with a waterfront winding from the Hudson River down to Sandy Hook Bay, along the Atlantic Ocean coast and up the Delaware Bay and River. The state, particularly its Atlantic coast, buffers two of America's most important cities, New York and Philadelphia, which therefore makes it an inviting enemy target in wartime. New Jersey was, in fact, harassed by sea raiders during both the War for Independence and the War of 1812. As technology progressed with the development of undersea combat potential in the late nineteenth and early twentieth centuries, war along the shore became more sophisticated and deadly. Interestingly, New Jersey had ironic—considering the ultimate outcome—and close connections with some of the more significant of these technical advances.

It could be said that the first submariner, or at least the first man to seriously posit the possibility of undersea warfare, was Cornelius Debrell. Debrell, who allegedly invented an "invisible eel to swim the haven at Dunkirk and sink all the shipping there," was a friend of King James I of England, who became the original owner of New Jersey in 1664, before he turned it over to his cronies Lords Berkeley and Carteret. If Debrell's "invisible eel" was ever built—and frankly, it does not seem likely—it did not survive the centuries and neither did plans for its construction. One writer, however, working from a contemporary description, thinks the proposed vessel may actually have been launched and functioned after a fashion, using "goatskins sewed together in the form of bags" that filled with water to submerge, after which it was supposedly propelled by underwater oars. An Englishman named Day allegedly constructed a submarine vessel "nearly one hundred years later" and successfully submerged it, "but never returned to the surface." Seventeenth-century Italian polymath Giovanni Borelli also drew up plans for a submersible vessel, although there is no evidence that it was ever actually built.[1]

It is generally believed that the first submarine that actually took to the water in an attempt to sink an enemy surface ship was invented by an American, David Bushnell, a Connecticut native and Yale student, who came up with his *Turtle* design in 1775. Bushnell's avocado-shaped wooden craft, outfitted with paddle propulsion (some mistakenly believe the vessel had a screw propeller) used water as ballast to enable it to rise and descend in the water. In the summer of 1776, Bushnell offered his one-man submarine design, which he had tested in the Connecticut River, to the American Revolutionary army, then defending New York from an impending British invasion. He proposed that his submarine could be used to sink British ships off the Manhattan shore, and a somewhat dubious George Washington decided to give it a try. The *Turtle* was supposed to approach an enemy ship while submerged and thus unseen and then drill a hole in its hull and attach an explosive device or mine, also invented by Bushnell. On September 7, 1776, the *Turtle*, operated by a volunteer, Sergeant Ezra Lee, reportedly did reach a British frigate, the *Eagle*, but the attempt at drilling failed, either due to the resistance of an iron band or the copper hull sheathing. Lee, unable to complete the mission, claimed he let the mine float away, where it exploded, "sending a column of water high in the air and creating consternation among the shipping in the harbor," although this assertion has been disputed.[2]

When the American army retreated out of New York, Bushnell and his submarine went with it. The craft made it as far as Fort Lee, New Jersey, where the ship it was being transported on was reportedly sunk, and the *Turtle* went down with it in the Hudson River somewhere around where the New Jersey end of the George Washington Bridge now stands. Bushnell went on to employ his floating mines during the 1777 Philadelphia campaign, drifting them down the Delaware toward the British fleet, creating "great consternation," although without any real damage to the enemy. He went on to become an officer in the Continental army and a medical doctor and eventually ended up settling in Georgia.

Robert Fulton, who launched the first steam ferry operation from Paulus Hook, in today's Jersey City, to Manhattan, also tried to develop a

David Bushnell's *Turtle* met its end in the Hudson River at Fort Lee, New Jersey. In September 1945, this model of the submarine, which had been owned by President Franklin D. Roosevelt, went on display at the New York Arts and Antiques Show. *Joseph Bilby.*

submersible vessel. Fulton's idea was far more complex than Bushnell's, and the result was the *Nautilus*, in which he allegedly "made numerous descents, and…covered fifty yards in a submerged run of seven minutes." In 1801, Fulton took his idea to France, where it was rejected by French admirals, who apparently thought the whole concept of undersea warfare unethical and reprehensible. Fulton, an equal opportunity salesman, went on to England, where the government allegedly paid him $75,000 to promise that he would not sell his plans to that country's enemies, although Fulton retained a caveat for his native United States. In the end, the idea came to nothing and Fulton returned to New York, where he worked on his ultimately successful steamboat.[3]

The Civil War provided a significant impetus to developing effective submarines. Although the Union navy essentially controlled the seas, despite havoc created by Confederate commerce raiders, the defenses of the Charleston, South Carolina harbor proved impenetrable by surface craft due to mines and obstructions. Submarines were seen as a viable alternative to the Charleston problem as well as useful in pushing up the James River toward Richmond and defending the blockading fleet against Confederate attacks.

The earliest of these submarine designs was the *Alligator*, an invention of French immigrant Brutus de Villeroi, who had been working on his "fish boat" project since the 1830s. He built a prototype, later named *Alligator Junior* by the press, in 1859. The *Alligator*, constructed in Philadelphia at naval request following rumors of Confederate ironclad warship development that resulted in the *Virginia*, née *Merrimac*, became the first official United States Navy submarine. It was tested in the Delaware River and, some say, in a southern New Jersey Delaware tributary and delivered to the navy in May 1862. The *Alligator* was towed to Norfolk and up the James River to City Point, Virginia, in June and, after the *Virginia* was scuttled, was repurposed to support the Union advance on Richmond. Although several potential missions were proposed for *Alligator*, all were scrapped for various reasons, including insufficient water depth and fear of capture, and the submarine was sent back to the Washington Navy Yard in July, where it underwent some more tests, the results of which were deemed "unsatisfactory."[4]

The inventor apparently used a version of Cornelius Debrell's underwater oars as propulsion for his thirty-foot-long vessel, but when the navy removed Villeroi, described as "a bit of a scoundrel," from the project in early 1862, they were replaced with a screw propeller, which increased the submarine's speed to four knots. The *Alligator* remained in Washington until March 1863, when Admiral Samuel DuPont requested that it be towed to Charleston for use in dismantling Confederate harbor defenses, but it was cut adrift in a heavy storm off Cape Hatteras on April 3 and sank. A NOAA search in 2005 failed to find its remains.[5]

The early submarine with the most direct connection to New Jersey, historically and in the present day, is the *Intelligent Whale*, developed by Massachusetts inventor Scoville Merriam. In November 1863, a group of New Jersey investors led by Colonel William Halsted, former commander of the First New Jersey Cavalry, funded the construction of the vessel in Newark, New Jersey. It featured a half-inch-thick wrought-iron hull and interior machinery, including valves, pumps and propulsion

The front of the *Alligator* coin, showing the submarine. *Joseph Bilby.*

The reverse of a coin commemorating the search for the Civil War submarine *Alligator*, tested in the Delaware River, which sank off North Carolina in 1863. The quote is from a letter by inventor Brutus de Villeroi to Abraham Lincoln. *Joseph Bilby.*

I propose to you a new arm of war, as formidable as it is economical. Submarine navigation, which has been sometimes attempted, but as all know without results, owing to want of suitable opportunities, is now a problematical thing no more.

Brutus de Villeroi in a letter to President Lincoln, 1862

Oliver Halsted, lobbyist and owner of the *Intelligent Whale*. *Library of Congress.*

equipment, and had a six-man crew, four of whom propelled it by hand cranking a four-bladed screw propeller. Like the *Alligator*, the *Whale* had a door in the bottom that could be opened to allow a diver to leave the submarine to remove obstructions or plant mines. The air pressure in the submarine exceeded the outside water pressure, thus allowing the diver to leave and return without the craft flooding.[6]

The owners hired well-known lobbyist Oliver S. "Pet" Halsted, a Newark attorney and relative of Colonel Halsted, to represent them in an effort to sell the submarine to the navy, and it was tested in Long Island Sound in August 1864, when it successfully submerged and then returned to the surface but was not prepared to demonstrate its other capabilities. The navy declined to purchase it due to fears regarding its seaworthiness, even after a more comprehensive test reported in the October 1864 issue of *Scientific American* magazine noted that "in all respects the vessel worked so completely that its success is undoubted."[7]

Unable to sell the *Whale*, Pet purchased it himself and, in March 1865, proposed a plan in which he, commissioned as a naval officer, would sail the submarine up the James River to Richmond. The war ended before Halsted could convince the Union high command to implement his idea, and so he brought the vessel to Newark and docked it at the Hewes and Phillips Machine Company on the Passaic River. He allegedly took it on pleasure cruises on the river while attempting unsuccessfully to sell it to Irish revolutionaries but eventually managed to convince the navy to purchase it in 1866. Transported to the Brooklyn Navy Yard, the *Intelligent Whale* was unsuccessfully tested there several years later and ended up as an ornament on the base commander's lawn. When the navy yard closed, the submarine was shipped to the Washington Navy Yard, where it remained until returning to New Jersey in 1999, where it currently resides on public display at the National Guard Militia Museum of New Jersey in Sea Girt.[8]

By the second half of the nineteenth century, a number of other underwater craft designs were in the works, here and abroad. Another would-be submariner, who had no direct connection to New Jersey but reportedly tested his vessel in waters that lapped the state's shore, was Josiah Hamilton Langdon Tuck, who sunk $16,000 into his *Peacemaker*, a thirty-foot-long "submarine torpedo boat" described in one newspaper account as looking like "a shark with a hole in its back" and tested it in the Hudson River between 1884 and 1886. During the test dive, it began to leak, making one of the crew "so hysterical that it was necessary to take up a hammer and threaten to brain him unless he became quiet and did as he was told." The inventor and crew escaped, but the *Peacemaker*'s brief career was over; Tuck's relatives, angered at what they perceived to be his squandering of family money, allegedly had him committed to an insane asylum, although there is no provenance for that claim.[9]

Enter John Philip Holland. Born on February 24, 1841, in Liscannor, County Clare, Ireland, Holland, the son of a local Coast Guard "riding

The *Intelligent Whale* on display at the National Guard Militia Museum of New Jersey in Sea Girt. *Joseph Bilby.*

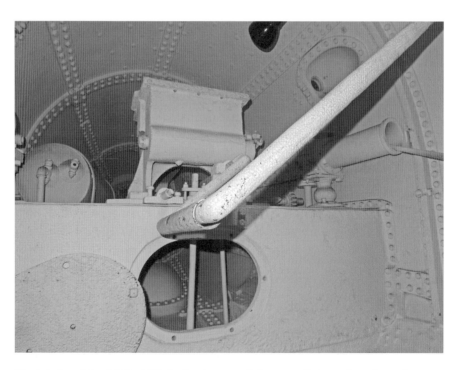

The interior of the *Intelligent Whale*, showing one of the manually turned cranks used to power the vessel. *Joseph Bilby.*

officer," had a childhood marked by intense intellectual curiosity that manifested itself in an interest in science and engineering. In 1862, Holland, who had joined the Irish Christian Brothers and was teaching school at the time, read about the combat between the ironclad warships *Monitor* and *Virginia* in Hampton Roads in the *Cork Examiner* newspaper, and it occurred to him, as a budding Irish nationalist, that ironclad vessels presented the possibility of ending British dominance of the seas. Needless to say, it also occurred to the British Admiralty, and shipbuilders were quickly put to work building ironclads to remedy the situation.[10]

More germane to our story, two years later, Holland apparently read an account of the *Intelligent Whale* tests conducted in the waters off New York, which motivated him to sketch out an idea for a submarine "so that it would be practical to live and work while completely submerged even in rough water." He used his self-taught engineering expertise to determine the correct shape of such a craft and the thickness of the metal needed to allow it to survive external water pressure to a depth of 250 feet. Holland could not, however, interest anyone in helping him bring his idea to reality.[11]

In 1872, Holland immigrated to Boston, Massachusetts, following his mother and his two brothers, both of whom were involved in the Irish revolutionary movement. While in Boston, he slipped on a winter ice slick, fell and broke his leg. To keep his mind busy while recovering, Holland dug out his old submarine plans and worked on improving the design, although he did not actively pursue the idea until three years later, when he was a teacher at Saint John's parochial school in Paterson, New Jersey, and proposed his concept to the United States Navy. The response was not encouraging. Admiral William T. Sampson concluded that "the project was practically impossible" as there was no way to "find in what direction to steer the boat under the water."[12]

Holland would not give up, however, and sought financial backing from the Irish revolutionary Fenian Brotherhood, an organization in which his brother Michael was an officer, by demonstrating his idea with a thirty-inch model at Coney Island, New York. The Fenians promised funding, and Holland contracted with the Albany Street Iron Works in New York City to build the framework of an actual undersea vessel to his specifications. The construction of the fourteen-foot boat was completed at a machine shop in Paterson, and it was test launched in the Passaic River in July 1878. The launch was a failure, much to the disappointment of an impromptu audience of millworkers who gathered to watch the scene with cheering and then "loud yells." The technical faults that caused the failure were remedied to a degree, and a subsequent test the following week was a bit more successful, with Holland reportedly remaining underwater for an hour before surfacing. According to one account, a man with "a powerful field glass" was observing the activities of Holland and his crew from a distance, and "it was said that he was an agent of the British Government."[13]

Not satisfied with his original boat, but considering it a valuable learning experience, Holland stripped it of reusable parts and proposed a second project to his backers. Despite some dissension in Fenian ranks about the feasibility of the plan, the Irishmen dipped deeper into their "Skirmishing Fund," an account dedicated to purchasing weaponry that could be used against the British, and asked Holland to come up with a design for a bigger submarine, which he did. Turning to another New York City ironworks, Holland launched the second generation of his invention in 1881 in the Morris Canal basin in the Hudson River at the Morris & Cummings Dredging Company's Jersey City dock. The ensuing tests made the newspapers, including the *New York Sun*, which dubbed the boat the *Fenian Ram*, a name that stuck. The publicity attracted a number of foreign visitors, including

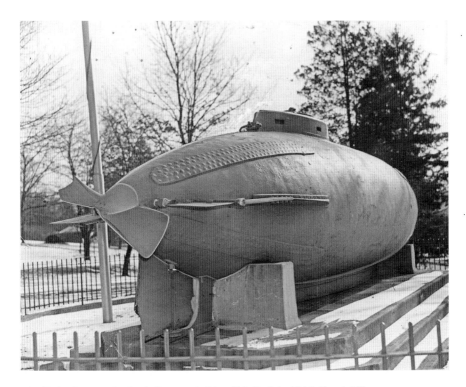

The *Fenian Ram* on display in Paterson's West Side Park in 1940. *Joseph Bilby.*

"Swedes, Russians, Italians, and Germans," who apparently concluded "that the project would never amount to anything." Holland believed that his uninvited guests had been conned by the English into disregarding what he perceived as a potential revolution in naval warfare.[14]

From his Jersey City base, Holland continued to experiment and make improvements on the *Fenian Ram* into 1882, cruising down past Staten Island and over to Brooklyn. He even experimented with launching primitive torpedo-like projectiles, which he obtained from *Monitor* designer John Ericsson, from a pneumatic gun in the bow. An initial problem resulting in the torpedoes flying out of the water—one almost hit a fisherman on a nearby dock—was resolved by further testing. Unfortunately, internal feuds in the Fenian organization led to one faction's decision to steal the *Fenian Ram*, as well as a smaller test model, and tow them to Connecticut. The smaller submarine sank on the way (it was later recovered), but the second-generation boat made it to its new home, where it was stored in an old shed, because the Fenians did not have the technical expertise to operate it, thus ending their "salt water enterprise." In the immediate aftermath of the boat-napping, Holland implied to an inquiring reporter that someone in the employ of "an English detective" might be responsible for the act. Later, however, he accused his former backers of "disgracing the name of Nationalists by laying themselves open to charges of stupid mismanagement, theft and swindling."[15]

By the time the *Fenian Ram* was kidnapped, Holland had resigned from the Irish Christian Brothers and his teaching position to become a professional submarine designer, but he suddenly no longer had a submarine. Within months, however, he had another vessel under construction at Thomas Gannon's Jersey City foundry. Holland launched the new submarine on May 30, 1883, and, when asked by a reporter, denied any affiliation with his Fenian former backers, stating that he was "into this on business principles" and was looking for a "capitalist" to fund his work. In subsequent years, Holland had a succession of partners and promoters, including Charles Morris of the Morris & Cummings Company, navy lieutenant William W. Kimball and army lieutenant Edmund L. Zalinski. Zalinski owned the Pneumatic Gun Company, a firm that manufactured a version of the "Dynamite Gun." Holland and Zalinski formed the Nautilus Submarine Boat Company, which produced yet another submarine, dubbed the Zalinski Boat, in Brooklyn. Unfortunately, the submarine was damaged and sank on launching, as did the company.[16]

When the navy appeared to become seriously interested in a submarine in the late 1880s, Holland, who had married a Paterson woman and moved to Newark, entered the competition, along with other hopeful designers. Unfortunately, the secretary of the navy changed his mind and spent the allocated money on surface ships. Close to poverty, the self-taught Irish inventor was rescued by his friend Charles Morris, who hired him as a draftsman. In 1893, naval interest in a submarine returned, and Holland, who had been awarded a patent for his design the previous year, again entered the competition, this time as an officer of the Holland Torpedo Boat Company.

After a series of design changes, Holland's final submarine, the sixth proposal of his career, which actually featured many of the characteristics of the earlier *Fenian Ram* but was fifty-three feet long, was built at the Crescent Shipyard in Elizabeth, New Jersey. *Holland VI* was launched on May 17, 1897, but damaged by a careless workman shortly afterward. It was finally declared ready after trial runs off Staten Island on March 18, 1898, followed by official tests before a Navy Board of Inspection in Raritan Bay shortly afterward. Naval indecision resulted in more tests, and when the Holland Torpedo Boat Company ran out of money, it merged with another entity, and the new business became known as the Electric Boat Company. Finally, in August 1900, the navy agreed to buy Holland's submarines from Electric Boat. The company manufactured or licensed construction in a number of other countries, including Britain, the Netherlands and Japan. Holland later left Electric Boat and unsuccessfully tried to found a submarine company of his own. Undeterred, he began to experiment with building a "flying machine." The underwater arms race he had begun continued at a rapid pace without him.[17]

As years passed, Holland's *Fenian Ram* became a well-traveled vessel on land. Initially stored in a lumber shed on the Mill River, the submarine reportedly lost its two-cylinder Brayton engine to power a forge. If so, it was returned, as it still has its engine aboard. In 1907, "Admiral" Jim Keegan, a Brooklyn boat captain, claimed that he had raised the *Fenian Ram* from the bottom of a Connecticut river, made it seaworthy and wanted, along with some friends, to sail it to Virginia to attend the exposition being held there to commemorate the 300th anniversary of the founding of the Jamestown colony. Keegan apparently had a vivid imagination and spun a tale for a reporter in which the submarine had once begun to sail from America to Britain "prepared to annihilate the British Navy" but somehow got turned around and ended up in Connecticut. The self-promoting Keegan, deemed a "suspicious character" by government officials, said that even if he did not get permission to attend the official Jamestown festivities, he would sail out

John P. Holland's successful submarine in U.S. Navy service in 1909. *Joseph Bilby.*

John P. Holland's submarine in 1902 shows the *Holland* along with the Russian battleship *Retvizan*, built by the Philadelphia shipbuilder Cramp and Sons. *Joseph Bilby.*

The *Fenian Ram* on display in Paterson's West Side Park in a 1964 postcard image. *Joseph Bilby.*

of Raritan Bay for Virginia flying "the green flag with the sunburst" of the Fenian movement. There is no evidence he ever did.[18]

In 1916, the *Fenian Ram* was transported to New York City, where it went on display in Madison Square Garden as a fundraiser for the survivors of the Easter Rebellion in Dublin. From there it was moved to the campus of the New York State Marine School at Clason Point in the South Bronx, where it remained until 1927, when it was purchased by Edward A. Brown of Paterson, who moved it back home to New Jersey, setting it up in the city's West Side Park, near the spot where Holland had conducted his first submarine test in the Passaic River many years before. It is currently on display in the Paterson museum.[19]

John P. Holland, the designer of the first truly successful submarine, lived out the rest of his years in New Jersey. He suffered a stroke in 1909, retired from public life in 1912 and died at his home at 38 Newton Street in Newark on August 12, 1914, shortly after the beginning of World War I, a conflict that, according to one obituary, would have enabled him to "observe the thorough test of his invention in the European war." That "thorough test" would also, ironically, bring undersea war to the shore of his adopted state.[20]

Chapter 2

THE CRUISE OF *U-151*

"THE OFFICER WAS SO POLITE THAT HE
ALMOST GOT ON OUR NERVES."

New Jersey was home to the first practical submarine, along with several of its less successful predecessors, but the state's acquaintance with underwater technology would turn sour in the twentieth century.

World War I began in Europe in 1914, following the assassination of Austro-Hungarian Archduke Franz Ferdinand and his wife, Sophie, by Serbian nationalist Gavrilo Princip in Sarajevo. When Austria-Hungary responded by invading Serbia, the series of alliances created in Europe in the decades before the assassination triggered a war of unprecedented and unthinkable dimensions—and horror.

Although President Woodrow Wilson, a former New Jersey governor, predicated his 1916 reelection campaign (headquartered in Asbury Park and West Long Branch) on the slogan "he kept us out of war" and gained a narrow victory over his Republican opponent Charles Evans Hughes, it was clear to many that war was indeed in the offing. The subsequent revelation of the contents of the disconcerting Zimmermann Telegram, in which Germany was encouraging Mexico into an alliance against the United States, pushed the country closer to intervening in the conflict, and the German declaration of the resumption of unrestricted submarine warfare on the open seas proved the final straw. The United States declared war on Germany on April 6, 1917.

With the notable exception of the Confederate commerce raider CSS *Tallahassee*—which captured and burned four merchant ships, the *A. Richards*,

Woodrow Wilson accepting the Democratic Party nomination for president at Sea Girt, New Jersey, in 1912. *Joseph Bilby.*

Carrie Estelle, William Bell and *Sarah A. Boyce*, off Sandy Hook on August 11, 1864—World War I brought serious hostilities to New Jersey waters for the first time since the War of 1812. Although New Jersey was never actually invaded by British forces during the earlier conflict, there are a number of stories of small local ships chased and captured and landings by enemy raiding parties from the British blockading fleet. Much of this information is sketchy. One source notes, "It appears they [Monmouth County militiamen] were engaged in one fight, called the battle at 'Brant Hill,' in which a British man-of-war, understood to be the 'Effervire,' carrying thirty-two guns, attempted to land at the mouth of the Squan River." According to the account, the British launched several small boats that attempted to capture and burn some ships in the river but were repelled by "volleys of musketry" from the militiamen. Local lore, recounted in a late nineteenth-century county history, has British sailors and marines from the blockading fleet landing near Barnegat Inlet to capture or destroy civilian vessels on several occasions, while local people watched the action from their rooftops in Forked River.[21]

Cape May, located at the tip of a peninsula, was in an unusually exposed position, subject to seaborne depredation from the Atlantic Ocean and Delaware Bay. In July 1812, the county board of freeholders brought two Revolutionary War cannons out of retirement, although Robert Holmes, the antiwar Federalist county treasurer, initially refused to pay for new gun carriages. In March 1813, the freeholders appropriated $300 for the purchase of "amputating instruments, gunpowder, and '100 weight of large buckshot'" for issue to the militia. Shortly afterward, money intended to fund the Dennis Creek Causeway was diverted to buy six hundred pounds of cannonballs, two kegs of powder and material to make artillery cartridges. As the British blockade began to materialize that year, Cape May militiamen built fortifications armed with log-barreled *faux* "cannons" at Goshen Creek.[22]

A British blockading squadron reached Delaware Bay in March 1813 and immediately conducted a loud, if largely ineffectual, bombardment of Lewes, Delaware, which spread panic on the Jersey shore side of the bay. People in inland New Jersey towns like Bridgeton feared incursions up the Cohansey or Maurice Rivers, which, fortunately for them, did not occur, although the British threatened Cape May residents with shelling their houses if they warned coastal trade ships of the enemy's location. In May, the British navy captured three sloops off the mouth of the Maurice River, and on May 30, a landing party reportedly skirmished with militiamen near Leesburg in Cumberland County. Raids and landings, as well as numerous false alarms, continued into late 1814, and nervous citizens drove their stock inland, buried their valuables in backyards or dropped them down wells when alarms were sounded. None of the raids, however, came close to the size of British and Loyalist incursions of the Revolutionary War era.[23]

Incidents at sea, mostly involving the blockading fleet and local merchant ships and occasional American privateers and British supply ships, also occurred off the New Jersey coast. In late 1813, the *New Jersey*, a small coastal trading ship out of May's Landing with a three-man crew, was captured by a British armed schooner off Cape May. The British put a small "prize crew" aboard, but the Americans managed to recapture their ship and bring it in to Somers Point, where two Englishmen in the prize crew "hired out in the vicinity" and a third, an Irishman, joined the American navy as a gunboat sailor. On July 4, 1813, an American subterfuge of hiding an armed party of sailors below decks on a local fishing boat, the *Yankee*, succeeded in capturing the British schooner *Eagle* off Sandy Hook. In another less than auspicious contest, a gunboat squadron from Philadelphia conducted a daring assault on a grounded sloop of the blockading squadron in Delaware Bay, but the fight ended when gunboat *121* was captured by the British and towed out to sea. The gunboat was later abandoned and washed ashore with the tide at Absecon, where the locals scrapped it for the iron and brass.[24]

As World War I raged on in Europe in the years following 1914, the United States remained officially neutral. Despite that, the country supported Allied forces fighting the Central Powers with military hardware sales. Major American firearms manufacturing companies, including Winchester and Remington, made rifles for the British, French and Russians, and other factories churned out ammunition and explosives from rifle cartridges to artillery shells and bombs. New Jersey was home to a significant number of those plants, and the Imperial Russian government established an ammunition testing facility in Lakehurst.

New Jersey had been a developing industrial state for many years, but according to one account, "the guns of Europe" were responsible for "the most intense industrialization in its history. The production of high explosives, textiles, steel, and ships rocketed to new heights. The Bureau of Statistics reported that expansion in manufacturing was 400 percent greater in 1916 than in any preceding year…The chemical industry in New Jersey sprang up almost overnight. Six factories for the production of aniline, formerly imported from Germany, were set up within the state, the most important at Kearny." Another source states that the state's overall industrial output "increased almost 300 percent between 1914 and 1919." By the time of the American entry into the war, New Jersey was the largest ammunition-producing state in the United States. Unfortunately, all of this expansion also made New Jersey a target.[25]

On July 30, 1916, the "Black Tom" ammunition pier on the Hudson River in Jersey City exploded. The force of the explosion broke windows all over Jersey City and Manhattan and damaged the Statue of Liberty and buildings on nearby Ellis Island. Damage in Jersey City alone was estimated at $1 million ($23 million in today's money). Large amounts of ammunition manufactured in the United States were shipped to the Allies in Europe through Jersey City and Hoboken, and it was believed German saboteurs had placed bombs on the pier.[26]

On January 11, 1917, the Kingsland ammunition plant in Lyndhurst, owned by the Canadian Car and Foundry Company, exploded. Switchboard operator Theresa "Tessie" McNamara was a heroine that day. She stayed at her switchboard, calling every building on the site, telling the workers to evacuate, even though exploding shells hit the building she was calling from, and all 1,700 workers escaped. An investigation into the Kingsland disaster concluded that it was also the result of sabotage.[27]

On the evening of January 12, 1917, 400,000 pounds of smokeless powder exploded at the Du Pont plant at Haskell, in Passaic County. Remarkably, only two employees were reported missing and twelve injured. The effects of the blast resonated for a 150-mile radius in New York and New Jersey. Shocks were reportedly felt as far away as Albany. Initially thought to be another case of sabotage, the Du Pont explosion was later determined to be the result of an accident. Other explosions would rock the state after American entry in the conflict, including a massive detonation on October 4, 1918, when the T.A. Gillespie Shell Loading Plant, located in the Morgan section of Sayreville, exploded, destroying the munitions manufacturing operation and setting off three days of intermittent blasts that destroyed more than three hundred buildings and created chaos in Sayreville and South Amboy. The populations of Sayreville, South Amboy and Perth Amboy were evacuated, and martial law was temporarily established by the state's home guard under orders from Governor Walter Edge. Over one hundred people are estimated to have perished. The Gillespie explosion was later determined to be an accident.[28]

Although postwar Germany contested the sabotage accusations regarding Black Tom and Kingsland, an international claims commission found the country guilty of sabotage in both explosions and awarded $50 million in damages to claimants. World War II intervened, but in 1953, the West German government, without ever admitting responsibility, agreed to pay the compensation. It was finally paid in 1979.[29]

With official American entry into the war, the Germans suddenly had a more effective weapon to use other than erratic sabotage efforts. In World War I, unlike in the War of 1812, the enemy would not land on the New Jersey shore but would concentrate on disrupting offshore shipping lanes. German submarine depredations along the state's coast during World War II are relatively well known, but the cruise of *U-151*, a German submarine that departed from the naval base at Kiel in April 1918 with the mission of attacking American East Coast shipping, is largely forgotten.

The crew of the German cargo submarine *Deutschland* in Baltimore in July 1916. *Joseph Bilby.*

U-151 was initially designed as the *Oldenburg*, one of a series of long-range cargo submarines, part of an early German plan to use mercantile undersea craft to evade the British naval blockade of Germany, sail to neutral ports and return with war supplies. The prototype of this class of submarine, the *Deutschland*, was launched in 1916 and left for America that June with a load of chemicals and dye. Once past the British blockade, *Deutschland* sailed on the surface, as it could travel faster that way. The submarine arrived at Baltimore on July 9 and returned to Germany "carrying 350 tons of rubber, 340 tons of nickel and 93 tons of tin." *Deutschland* sailed back to the United States in October, arriving at New London, Connecticut, with another cargo of chemicals and dye and left for Germany with a load of nickel and copper. The submarine accidentally rammed and sank the steamship *A.H. Scott* off Block Island, however, and had to return to New London for repairs, giving American authorities a good opportunity to take a look at its construction.[30]

In February 1917, *Deutschland*, along with other cargo submarines in the process of construction, was turned over to the *Kaiserliche Marine* to be armed and used in long-range military missions as "cruiser" submarines. *U-151* was equipped with two torpedo tubes, carried eighteen torpedoes and mounted two fifteen-centimeter deck guns with 1,672 rounds of ammunition and had a cruising range of twenty-five thousand nautical miles.[31]

Another German submarine, *U-53*, visited the United States in 1916. *U-53*, built in 1914, was an improved version of the standard German prewar submarine. *U-53* sailed into the Newport, Rhode Island harbor with permission from American authorities on October 7. The commander stated that he was paying a courtesy visit, and American naval officers were allowed to inspect his vessel in detail. After leaving Newport the following day, saluting the American navy on the way out, *U-53* went on to sink one Dutch, one Norwegian and three British merchant ships within hours. In a case of historical irony, *U-53* would go on to be the first German submarine to sink a United States Navy ship, the destroyer *Jacob Jones*, torpedoed off Queenstown, Ireland, on December 6, 1917.[32]

Even before entering the European conflict, the United States Navy had been concerned with the possibility of German submarines penetrating American waters "in case of war," a fear no doubt exacerbated by the voyages of the *Deutschland* and *U-53*. One postwar navy source noted that although *Deutschland*'s trips were apparently "purely commercial voyages," *U-53*'s visit possessed "the character of a path-finding expedition." In February 1917, the navy began to construct a "heavy wire" steel antisubmarine net, based on a British model that had proved successful, at Brooklyn Navy Yard. The net was intended to stretch between "Sandy Hook and Rockaway Point, crossing the three main channels—the Swash, the Old Main Ship Channel, and the Ambrose Channel," to protect New York Harbor. It was designed to only be used at night, with "torpedo boat destroyers, aeroplanes and a mosquito fleet which will be created for the purpose" taking up the antisubmarine task during daylight hours.[33]

The navy was also concerned about the efficacy of the existing New York Harbor defenses, including the large army base at Fort Hancock on Sandy Hook. The forts were equipped with long-range large-caliber artillery pieces and heavy mortars intended "to deal chiefly with battleships," not with submarines. The navy recommended that the harbor defenses be increased by adding "numerous searchlights and guns of small caliber," which it deemed better able to deal with relatively small craft. Still, whether for morale purposes or mere training, the big guns of Fort Hancock were fired on a fairly regular basis during the war, and the shock waves shattered windowpanes at nearby Sandy Hook Lighthouse regularly.[34]

By early 1918, Americans were well aware of the dangers of crossing the Atlantic, as submarine warfare had become a significant aspect of the German war effort. The *Tuscania*, a British Cunard ocean liner serving as an American troop transport that left Hoboken for France on January 24, 1918, with 2,013 American soldiers and a crew of 384 aboard, was torpedoed off the coast of Northern Ireland on February 5, with a loss of over 200 soldiers

Fort Hancock at Sandy Hook, New Jersey. *Joseph Bilby.*

and crewmen. The *Tuscania* was the first American troop ship lost to an enemy submarine attack, and the disaster received considerable attention in the press, where it was described as a "great shock."[35]

Assistant Secretary of the Navy Franklin D. Roosevelt tried to boost domestic morale with an address at Harvard University in January. Roosevelt posited, in essence, that British ship tonnage losses in the Napoleonic Wars surpassed those in the current conflict and that the "American people should not be discouraged" by submarine warfare. The newspapers passed Roosevelt's words of wisdom on to the general public: "Never had there been an invention of war, he said, that sooner or later had not an answer. The armor-piercing shell was the answer to the armor plate. The answer to the submarine peril, he said, was the building of more destroyers, chasers, patrol boats and a great merchant marine." In 1916, the prescient Roosevelt had been responsible for the development and manufacture of wooden-hulled sub chaser boats. As with most things in the industrial age, there is a New Jersey connection. Elco Manufacturing, located at Avenue A and North Street in Bayonne, made 722 of these small craft during World War I.[36]

In February 1918, the United States government announced that it had a solution to the submarine problem in the works, as Henry Ford had traveled to Washington, conferred with administration officials and agreed to produce "submarine killer" vessels. These ships, larger than those of the "mosquito fleet" sub chasers intended solely to protect the coast, were designed to escort convoys but also provide coastal defense. Ford proposed producing these ships by expanding his River Rouge plant near Detroit and also by adding a new plant to be built "on an eighty-acre tract of land on the Lincoln Highway between Newark, N.J. and New York City." The New Jersey factory, proposed as a "duplicate of the River Rouge shipbuilding plant now in course of erection," was apparently never completed, although Ford did establish an auto plant in Kearny, later closed when a more modern facility was constructed in Edgewater, New Jersey, in 1930.[37]

The new antisubmarine ships were dubbed Eagle boats, and by April, navy secretary Josephus Daniels was out in River Rouge inspecting the facility and the first of the Ford ships, allegedly nearing completion. By June 6, with no Eagle boats yet launched, a newspaper account reported that "Herculean efforts will be made to speed up production of Henry Ford's eagle boats for use in coast patrol service against submarines." On July 11, the first Ford-made antisubmarine ship, officially dubbed *Eagle No. 1*, was "dropped into the water" at River Rouge. None of the Eagle class ships ever saw combat in World War I.[38]

U-151 reached American waters, undisturbed by the uncompleted Eagle boats, in early May 1918, and the sub's skipper, Kapitan Van Nostitiz und Janckendorf, immediately set about performing his initial mission, laying mines, beginning off the North Carolina coast near Currituck Sound. Moving north, *U-151* laid more mines off the Chesapeake Bay inlets at Cape Henry and Cape Charles. Van Nostitiz used American lighthouses and lightships to assist in determining his position and was surprised at the lack of war preparation he witnessed.[39]

Naval authorities were aware that an enemy submarine might be sailing toward the United States, since they had received several reports from the British steamships *Port Said* and *Huntress*, which had been fired at with torpedoes and missed by *U-151* as it traveled to the East Coast. In response, the navy issued an order instructing merchant ships to take care that "no lights should be carried, except as may be necessary to avoid collision, and paravanes [a British-developed gliding underwater mine-sweeping device that would cut mine cables or explode them harmlessly] should be used when practicable and feasible." They did not, however, call for dim-out or blackout conditions to be established in shore communities.[40]

Several World War I–era paravanes at Fort Meagher in County Cork, Ireland. Meagher, called Fort Camden under the British in World War I, is one of the defensive forts at the entrance to Cork Harbor. *Damian Shiels.*

Opposite, top: A *U-151*-type submarine in port in Germany. *Joseph Bilby.*

Opposite, bottom: The type of mine sowed by German submarines in World War I. *Joseph Bilby.*

In the process of moving north, *U-151* attacked three sailing schooners: the *Hattie Dunn*, the *Hauppauge* and the *Edna*. The Germans did not use torpedoes in these attacks but surfaced and halted the vessels to sink them by planting explosives with time fuses after ordering the crews to abandon ship. While the *Hattie Dunn* and the *Hauppauge* quickly sank, the *Edna*, attacked near the Winter Quarter Shoals lightship off Chincoteague, Virginia, remained afloat, albeit damaged, and was later towed into Philadelphia, alerting naval authorities to the presence of a German submarine close to shore.[41]

Lieutenant Fredrick Koerner, the German "boarding officer," politely asked each ship's captain for his vessel's papers and then provided receipts written in German in exchange before ordering the crews to leave in lifeboats and row to his submarine, which, since it had been designed as a mercantile ship, had room for the twenty-three merchant seamen taken as captives, to

A schooner destroyed by *U-151* explosive charges. *From* Raiders of the Deep.

be held until *U-151*'s mining operation was over so as not to alarm local authorities to the submarine's activities. Adolph Lewis, the African American first mate of the *Edna*, later recalled, "The food was very poor. The best we got was canned brown bread, tea and coffee without sugar and poor stews." He remembered that most of the crew spoke English and that there was a "canteen" where the prisoners could buy cigars and cigarettes, the Germans taking payment in American money they wished to bring home as souvenirs. The lifeboats from the three vessels were lashed down to *U-151*'s deck. The German courtesy in evacuating the crews before sinking their ships was presumably due to the fact that they were operating under "Cruiser Rules," a long-established maritime warfare code of conduct specifying that unarmed merchant vessels should be warned before they were attacked.[42]

Lieutenant Fredrick Koerner, *U-151*'s "boarding officer." *From* Raiders of the Deep.

In a postwar interview, Lieutenant Koerner expressed a sentimental regret at sinking the schooners: "We blew up the *Hauppauge* with TNT. Masts and spars and deck rails sailed high in the air. What a sight! Wonderful in a way, but one to make a sailor's heart grow heavy. In this time of ocean liners, a fine, trimly rigged schooner is one of the last reminders of the picturesque old days."[43]

By the end of the month, *U-151* and its prisoners reached New Jersey waters, laying more mines off Delaware Bay, near Cape May and Cape

Henlopen, and then sailing north to cut telegraph cables to Central America and Europe sixty miles southeast of Sandy Hook on May 28. Koerner recalled that "our cable cutting mechanism dragged along the bottom. We waited patiently for a bite, that feel of the line that would indicate that we had caught hold of the cable. Then we cut it. Every time a ship hove into sight, we would close our hatches and submerge."[44]

One of the Delaware Bay mines would damage the oil tanker *Herbert L. Pratt*, sailing from Mexico to Philadelphia, on June 3. The *Pratt* was beached, and patrol boats from Cape May and Cape Henlopen came to its aid. The crew was conveyed without loss to Lewes, Delaware, and the ship was repaired and refloated by June 5. The worst was yet to come, however, as the German submarine turned south again after cable cutting. On June 2, 1918, *U-151* sank six ships off New Jersey on a day locals recalled as "Black Sunday."[45]

The doleful day began when *U-151* stopped the schooner *Isabel B. Wiley*, a sailing ship that headed out past Sandy Hook on the afternoon of June 1 on its way to Newport News, Virginia, and Montevideo, Uruguay. Just before 8:00 a.m. on June 2, Captain Thom I. Thomassen spotted a "suspicious looking object" approaching his ship and soon saw that it was flying "a small German naval ensign." The submarine, moving on the surface, fired a warning shot from one of its deck guns, and Thomassen stopped his ship and "hauled down the jibs." While he was waiting for Germans to board, *U-151* fired a shot at the *Winneconne*, a ship steaming along nearby en route from Newport News, Virginia, to Providence, Rhode Island, with a load of coal. The submarine fired yet another shot at the *Winneconne* to halt it as Germans boarded the *Wiley*, placed explosive devices and told the crew to get off.[46]

The Germans kept the *Isabel B. Wiley* afloat while the crew left and then took supplies aboard the submarine and transferred some of their prisoners to the lifeboats that had been lashed to the deck, turning them over to custody of the *Wiley*'s captain. With all the mines laid, there was no longer a reason to keep them captive. Kapitan Van Nostitiz gave Thomassen some fresh water and told him to head for shore. The American captain had the only powerboat, so he went to look for help while the others remained in place. He eventually came into contact with the Ward steam liner *Mexico*, heading south, which picked up the stranded sailors. The *Mexico* contacted the *Santiago*, heading north, and transferred the men to that ship, which brought them into New York Harbor.

When Captain Waldemar Knudson of the *Winneconne* asked the Germans why they were boarding his ship and planting bombs to sink it, Lieutenant

Koerner "said he was sorry, but war was war and England was to blame." Another officer told the crewmen "you men take to your boats...I got to sink you." They subsequently rowed up to the submarine, took on some additional freed prisoners and then "pulled for dear life to the westward" until they were "picked up by the *San Sabo*, about 25 miles southeast of Barnegat," the following morning.[47]

A decade later, Koerner nostalgically remembered the parting as perhaps more cheerful than it was, recalling that the prisoners "left in single file, each with a cordial goodbye," and the captains "seemed loathe to leave. They seemed to have grown fond of the company of the undersea privateers among whom they had been thrown so strangely." He recalled that "we shook hands...and the captains got into their lifeboat, vowing that they would that night drink a stein of beer to our health." There was to be no stein of beer that night. First Mate Lewis of the *Edna* recalled that he and his fellow survivors traveled into a heavy rainstorm and that "our boats began to fill...we seemed doomed... [and] we talked to one another to keep from going mad." After being rescued and arriving in New York, Lewis remembered, "We beat it for a first class American restaurant and a first-class feather bed."[48]

It was about noon when the crew of the *Winneconne* pulled away to the west and heard gunfire in the distance. Those shots were aimed at stopping the schooner *Jacob M. Haskell*, heading from Norfolk, Virginia, to Boston. After firing, *U-151* hoisted a flag displaying "the international signal 'Abandon Ship.'" As the *Haskell* crew climbed into lifeboats, a boat from the submarine pulled alongside with a party that boarded, demanded the ship's papers and planted explosive devices. Captain W.H. Davis recalled later, "The men [Germans] went about their work in a business-like manner; the officer [Koerner] was so polite that he almost got on our nerves. Each seaman was armed with two automatic revolvers [sic] and a long vicious-looking knife." As the *Haskell*'s cook was leaving, he suggested to Lieutenant Koerner that he might want to take some of the schooner's food on board, but the German responded, "We don't want your food; we have plenty food of our own. We don't want your lives either; we want your ships. Now get away from here; you have three minutes before the ship goes down." As the *Haskell* exploded and sank, the Americans rowed past *U-151*, and a German officer called out, "Good luck. The New Jersey coast is just 40 miles away. Better go there." The crewmen never made it to the coast, as they were picked up by the steamer *Grecian*.[49]

After sinking the three ships, *U-151* moved east and then southwest, where it encountered the schooner *Edward H. Cole*, on its way from Norfolk,

A German image of passengers and crew leaving a ship about to be sunk by a submarine. *Joseph Bilby.*

Virginia, to Portland, Maine, with a load of coal, south of Barnegat Light in midafternoon. The German submarine did not fire warning shots this time but sailed directly toward the schooner. First Mate Robert Lathigee of the *Cole* recalled that "nobody was thinking of U-boats when somebody spots something black about a mile to port." As *U-151* closed on the *Cole*, Lathigee assumed it was an American navy vessel, until it stopped 150 feet away and the crew was ordered to abandon ship. Following their by now standard pattern, the Germans boarded the *Cole* and planted explosive devices. The ship's captain, H.G. Newcombe, recalled that his ship went down "about 16 minutes after we left." He and his crew were picked up at 8:00 p.m. by the American steamer *Bristol*, which took them into New York.[50]

The crewmen of the *Cole* were about four miles away from their ship when the steamer *Texel*, on its way from Puerto Rico to New York City, "hove in sight." They heard the German warning shots and saw the *Texel* turn in an attempt to elude the submarine. Captain K.B. Lowry of the *Texel* had instructions from the navy on board and tried to follow them in an attempt to escape *U-151* but failed and eventually stopped. Before the Germans boarded, the captain tossed the navy instructions overboard. Lieutenant Koerner allowed Lowry to watch "the placing of the bombs" that would sink his ship. In the process, Koerner remarked, "I know how to do this, I have been in the business for four years."[51]

As the Americans hurriedly left the *Texel*, one man, Frank Ryan of New York City, took a moment to rescue the ship's mascot, a Maltese cat. Within eight minutes of Lowry and his crew departing *Texel*, the ship exploded and sank, and the crewmen "shaped our course for Absecon and pulled away." It was a rough trip. William Laufer of Millington, New Jersey, recalled, "We had a few biscuits and a little water and, among the 36 men, these did not go far. In the broiling sun yesterday we suffered terribly, and several of the men, unused to the exposure, showed signs of being overcome. To those we doled out the water as sparingly a possible, the hardiest of us depriving ourselves." The *Texel* crew finally rowed their way to Atlantic City inlet, from where they were taken to a hotel in the city by local police and instructed not to speak to reporters.[52]

Shortly afterward, however, Captain Lowry provided a detailed account of the attack to government interviewers. He noted that the captain of *U-151* "was about 5 feet 8 inches; probably weighed about 200 pounds, stocky build; he had light hair, wore a mustache and Van Dyke beard; he was about 40 years of age and wore a Navy uniform with long overcoat. His rank was indicated by two gold stripes on his sleeve, slightly about his wrist. He spoke English."[53]

U-151's final Black Sunday victim was the steamer *Carolina*, heading from San Juan, Puerto Rico, to New York. The *Carolina* was the largest ship attacked that day, with 218 passengers, a crew of 117 and a hold full of sugar. Captain Barber of the *Carolina* was aware of the presence of a German submarine in the area, as his wireless operator, E.W. "Sparks" Vogel, of Paterson, New Jersey, had received a message prior to 6:00 p.m. advising that the *Isabel B. Wiley* had been attacked earlier in the day thirteen miles to the north. He ordered all of the ship's lights doused and started to sail as fast as he could go due west.

As with Lowry, Barber's evasive tactics did not help him. *U-151* surfaced and fired three warning shots, after which the captain, "realizing the uselessness of trying to escape," had Vogel send a wireless SOS message to Cape May indicating that he was under attack by a German submarine. *Carolina* halted and *U-151* approached, flying a flag marked with the "A.B." signal, meaning abandon ship as soon as possible. Barber complied, evacuating women and children first and destroying "all the secret and confidential papers" before departing himself. Although a passenger said that the departure was orderly, Lieutenant Koerner remembered, "A great wailing of women's voices... praying and pleading."[54]

Passengers leave the *Carolina*. *From* Raiders of the Deep.

Sinking of the *Carolina*. *From* Raiders of the Deep.

All personnel were in lifeboats by 6:30 p.m., and then Barber was "ordered by the submarine commander, both in English and by signals with the hand, to make for shore." He later recalled that he "collected all the boats near me and moored them head and stern one to the other. Being eventually joined by all the boats except the motor lifeboat and lifeboat No. 5, we pulled to the westward and out of the line of gunfire as much as we possibly could." Instead of planting explosives on board, *U-151* tried to sink the *Carolina* with a torpedo, which missed, and then used three well-placed rounds of deck gunfire. "Great clouds of fire and steam arose as *Carolina* went down," flags flying.[55]

Although the chain of lifeboats got separated in a storm, many of *Carolina*'s passengers and crew were saved by the schooner *Eva B. Douglas*, which brought them into Barnegat Inlet on the morning of June 3. Others were picked up and brought into various ports, and one lifeboat rowed to the Atlantic City beach. The heroine of that craft was a young woman named Lillian Dickinson, who had served as an ambulance driver in France. Another passenger recalled that "although we tried to dissuade her, that gallant girl insisted on taking her regular turn at the oars. It was one hour on and one hour off. Just to look at this beautiful young woman keeping pace with whatever the men did was enough to buoy our spirits and keep us going. She set the pace and the example." Not all survived, however. The *Carolina* sinking resulted in the only casualties suffered on Black Sunday, when a lifeboat with eight passengers and five crewmen capsized during the overnight storm. The boat was eventually found by a passing Danish steamship, but its thirteen occupants were lost forever.[56]

On June 4, the *Santiago*, carrying Captain John Sweeney of the *Hauppage*, along with nine members of his crew and the crew of the *Wiley*, reached New York Harbor. Once journalists became aware that the ship carried survivors of a German submarine attack off the New Jersey coast, they flocked to interview them. Sweeney and his men had spent some time as prisoners aboard *U-151* and were particularly talkative. They opined to reporters that the submarine had headed back to Germany because "the commander said that he had only fuel and provisions for ten weeks, and had left Kiel on the night of April 19. The loaves of black bread, hard as flint, which were served out to the American crews during their eight days on board were all stamped April 19."[57]

There was apparently a good deal of fraternization between the crew of *U-151* and the captured sailors from the *Hauppauge*, who were "unanimous in their opinion that the crew of the German submarine, about seventy-six, were confident they were going to win the war and that nothing America could do with the allies would stop them," although they were concerned that their families in Germany could not get enough to eat. After the sailors had filled in the reporters, navy officials arrived and instructed them not to talk about their experience, but the story was out.[58]

As word of the attacks spread, in newspapers and through the coastal grapevine, the reaction from Washington was a mix of positive spin and assurance that the situation was under control. Naval officials claimed that "Germany, by striking with her submarines at the very doors of America, has admitted to the world that the American army will turn the tide against her on the battlefield of France." They also asserted that "the American anti-submarine force in home waters was able to meet the attack. All along the coast line, naval flying boats, submarine chasers and numberless other naval craft immediately got into action." Rumors spread that there was more than one submarine operating off New Jersey, which accelerated the defensive actions. As a precaution, the Port of New York was closed to outgoing shipping on June 4.[59]

In Asbury Park, the raid of *U-151* was considered more of a tourist event than a threat. Under the headline "Asbury interested but in no way alarmed," the *Asbury Park Press* noted that local people and vacationers were

> *evincing considerable interest in the work of the navy to protect the shores from U-boat activities. Yesterday afternoon many persons flocked to the beach to witness the patrol work of the airplanes and a dirigible which flew low over the surface of the water keeping a watchful eye open for a submarine. Rather than having the effect of alarming the city, the possibility of a Hun raid appearing off the coast has aroused the keenest interest and is resulting in many going to the boardwalk in the hope of seeing a chance encounter between the naval airplane and sub.*[60]

U-151 left New Jersey waters as defensive measures were activated but continued to attack American coastal shipping through July 2, when it returned to Kiel, arriving home on July 20 with, according to the United States Navy, an overall score of twenty-two ship sinkings (Koerner claimed twenty-three) totaling fifty-two thousand gross tons. Inspired by *U-151*'s success, several other German submarines sailed to the East Coast of the United States in 1918, but none operated as boldly off the New Jersey shore as *U-151* did, and their success was variable.[61]

The tempo of war in New Jersey waters significantly diminished after Black Sunday, but danger still lurked and casualties would mount. On

August 13, *U-117* torpedoed the cargo ship *Frederick R. Kellogg* twelve miles north of Barnegat Light and five miles offshore. Although seven crewmen were killed, the ship was salvaged. Heading south, *U-117* stopped the fishing schooner *Dorothy B. Barrett* about twenty miles off Cape May, allowed the crew to leave in lifeboats and then sank the ship. *U-151*'s mines lingered as well. The steamer *San Saba*, sailing from New York to Tampa, hit a mine a considerable distance at sea from Barnegat on October 3 and sank. There were only three survivors from the crew of thirty-four. The Cuban steamship *Chaparra*, bound from Havana to New York with a load of sugar, hit a mine ten and a half miles southeast of Barncgat light on October 27, 1918. Captain Jose Vinolas successfully evacuated his eleven-man crew and brought them in to the Barnegat Inlet Coast Guard Station.[62]

Three weeks after *U-151*'s rampage, Navy Secretary Daniels, perhaps trying to convince the public that he was on top of all possibilities and in charge of the situation, proposed a thousand-dollar reward to anyone providing "information leading to the discovery of enemy submarine bases on this side of the Atlantic." Daniels perhaps had second thoughts, as the offer was subsequently qualified by adding that "it does not signal official belief in the existence of a single German base anywhere from the broken coast of Maine to the Florida Keys to the lagoons of Louisiana." Daniels, who gave as his source "some of the admirals," said that "the enemy might have agents, somewhere along the coast or elsewhere, engaged in furnishing him with supplies," and that his proclamation was due to his "taking no chances." The offer was also open to "the West Indies, Mexico and the Spanish Main," as "a thousand dollars is a good deal of money to a dusky fisherman or a beachcomber." Unsurprisingly, no one ever collected.[63]

In the wake of the sinkings, coastal guardians got jittery. On the evening of August 1, 1918, a Captain Mollere of the Twenty-second United States Infantry Regiment, stationed on Sandy Hook, reported that "two small skiffs had been sent out from a submarine, presumably to make a landing off the Sandy Hook General Supply Depot." Major Fields, officer of the day at Fort Hancock, later confirmed that the invaders were actually New Jersey fishermen. Later that night, the Supply Depot guards heard four shots, one of which struck the sand near them. After an additional six shots, which seemed to come from the ocean, they returned fire and asked Fort Hancock personnel to train a searchlight on the water. By the time the light appeared, the shooting had stopped, and its source was never determined. Although *U-117* would have passed by that section of the New Jersey shore out at sea in early August, it is extremely unlikely that the firing came from a German submarine, as it was not within the mission of the U-boats, ordered to sow mines and sink merchant ships, to fire rifle shots randomly at the shore.[64]

At 2:00 a.m. on October 1, 1918, one of Franklin D. Roosevelt's wooden-hulled United States Navy "sub chasers" was sunk off the New Jersey coast, losing two men, the last casualties of the state's first coastal submarine war. A U-boat was not responsible, however, as the navy craft had crashed into an American oil tanker, the *H.G. Wells*.[65]

What was the mission of *U-151* and other German submarines along the East Coast in 1918? Initial public statements claimed, "The attack upon American shipping almost at the very entrance of New York Harbor is taken to mean that Germany has at last inaugurated a submarine campaign to break up transport of troops to France." The United States Navy's final assessment of the voyage of *U-151* was, however, that while the interruption of convoys was an objective of the German U-boat fleet, the main submarine missions along the New Jersey coast were probably to lay mines and destroy vessels that they happened upon in order to undermine American civilian morale. This would, they hoped, in turn create a demand to withdraw American naval vessels from convoy escort and mine-laying duty in the North Sea back to defend the homeland coast, an assessment that seems likely, and which Lieutenant Koerner's testimony seems to support.[66]

After the war, *U-151* was taken as a war prize by the American navy and subsequently used as a target in bombing tests off the coast of Virginia on July 6, 1921, where it was sunk and remains to this day. Some locations of the wrecked remnants of *U-151*'s Sunday rampage are known as well. The *Carolina* is in 240 feet of water sixty-five miles due east of Atlantic City, where it was discovered in 1995 by diver John Chatterton, who would subsequently be associated with the most famous New Jersey shore submarine story. Chatterton has retrieved artifacts from the wreck that prove beyond a doubt that it is the *Carolina*. Other divers claim to have found the wreck of the *Texel* some sixty miles off the New Jersey coast and the *Wiley* fifty-eight miles due east of Barnegat Inlet. The wreck of the *Winneconne* was allegedly discovered around the year 2000, in 210 feet of water some eighty miles off Manasquan Inlet, although there is no physical identification provenance and the location seems suspect considering the proximity of that ship to the *Wiley* when it was sunk.[67]

Following the allied victory in November 1918, Governor Walter Edge of New Jersey, who would govern the state once again in the final years of World War II, struck a positive note. Edge boasted that his wartime tenure had, in all, been a positive experience for New Jersey, as it had been the occasion of unparalleled prosperity and had been responsible for "the inauguration of the State highway system, the Delaware River Bridge, and the Hudson River tunnels," all of which was quite true.[68]

Reflecting on his cruise some years later, Lieutenant Koerner, dwelling more on the military consequences of the conflict, said: "For those who can see into the future surely this is a warning of what later wars may bring. For the day will come when submarines will think no more of a voyage across the Atlantic than they do now of a raid across the North Sea…America's isolation is now a thing of the past."[69]

Koerner's musing proved prophetic. The next time German submarines arrived at the Jersey shore, twenty-four years after the cruise of *U-151*, their commanders would not be "so polite," the duration of their stay would be longer and the ocean and beaches would be strewn with oil, wreckage and bodies.

Chapter 3

WORLD WAR II

"WHEN I CAME TO I SAW THE WHOLE AFT AFLAME"

It was just after midnight on February 27, 1942, and the *R.P. Resor*, built at the Kearny, New Jersey shipyard in 1935 and named after Standard Oil treasurer Rubin Perry Resor, was sailing up the New Jersey coast about eighteen miles east of Barnegat Light. The Standard Oil tanker had shipped out of Baytown, Texas, with a cargo of 78,729 barrels of fuel oil to deliver to Fall River, Massachusetts. The *Resor* was following all U.S. Navy anti-submarine instructions, sailing in a zigzag manner with all lights out, radios off and a deck gun manned by navy armed guards day and night. John K. Forsdal, the ship's lookout, spotted some flashing lights between his ship and the shore, thought they indicated a fishing boat and alerted the captain to avoid a possible collision. Moments later, however, his vessel experienced a "searing explosion" as a German torpedo hit it amidships. The *Resor* was immediately engulfed in flames, and the resulting "spectacular towering fire" was visible from as far away as the Asbury Park boardwalk and the Belmar Coast Guard Station. Forsdal recalled, "I was lifted off the deck and I was unconscious for a second or two. When I came to I saw the whole aft aflame." He looked toward shore in the blazing light and saw the silhouette of a surfaced German U-boat.[70]

Forsdal dropped a life raft overboard, jumped into the oily water and grasped onto it with fellow crewman Clarence E. Armstrong. When they saw a Coast Guard boat, *CG 4344*, approaching several hours later, both men screamed for help. The Coast Guard managed to save Forsdal, although it took several men to pull his oil-coated body aboard, but Armstrong slipped under the waves. The *CG 4344* sailed as close as it could to the burning *Resor* and later reported that "the sides of the ship were white hot" and that the paint on the side of *CG 4344* was "blistered" by the intense heat.[71]

The Neutrality Act of 1936 prohibited arming merchant ships, but with war raging in Europe, German submarines and aircraft roaming the Atlantic and the importance of sea lanes to logistical support, Section 6 of the Neutrality Act, the specific legislation barring the practice, was repealed on November 17, 1941, and the navy began to arm these vessels with four- and five-inch deck guns and machine guns, many of which were installed in Hoboken, New Jersey, and assigned and trained crews, known as armed guards, to man the weapons. Navy coxswain Daniel Hey, a member of the

Workers in Hoboken, New Jersey, constructing gun mounts to place on merchant ships for the armed guard program in November 1941. *Joseph Bilby.*

Opposite, top: A deck gun being mounted on a merchant ship in Hoboken, New Jersey, as part of the armed guard program in November 1941. *Joseph Bilby.*

Opposite, bottom: Daniel Hey (left) and John K. Forsdal (right) at the Manasquan, New Jersey Coast Guard Station. They were the only survivors of the sinking of the *R.P. Resor. Joseph Bilby.*

eight-man armed guard detailed to man the deck gun on *Resor*, was sleeping in his bunk when the torpedo hit. He ran up to the deck, where he and three other men tried and failed to launch a lifeboat through the flames. They all jumped overboard, and Hey swam for an hour and a half before he was picked up, alone, by a rescue boat. Forsdal and Hey, their bodies covered with heavy oil, were the only survivors of the *R.P. Resor*. Over the next day, "one body was picked up at sea by a Coast Guard patrol crew and three other bodies were reported sighted floating in the oil clogged waters off Manasquan Inlet." Forty-one merchant seamen and eight navy personnel on board failed to survive the attack.[72]

A newspaper reported that the resulting "blanket of fire, which turned into a pall of smoke as the morning and afternoon wore on, could be seen for hours afterward." Forsdal and Hey were brought to the Manasquan Coast Guard Station, where the oil was washed off their bodies and they posed for a photo—the two luckiest men on the New Jersey shore that day. Later in the afternoon, journalists took a boat out from Manasquan to photograph the smoldering wreck, which was still afloat, and spotted three more bodies floating in the oil slick around the *Resor*. The hulk

The *R.P. Resor* burned for some time before sinking. *Joseph Bilby.*

floated on fire for two days, providing a tourist attraction for crowds on area boardwalks before an attempt to tow it to shore for salvage failed. The sinking of the *R.P. Resor* brought World War II home dramatically to Monmouth County and New Jersey.

It was no real surprise to those in charge of defending the coast. In December 1941, Admiral Adolphus Andrews, commander of the North Atlantic Naval Coastal Frontier, extending from the Canadian border to North Carolina and stretching roughly two hundred miles to sea, knew the Germans were coming. (In February 1942, his command would be extended to Jacksonville, Florida, and be renamed the Eastern Sea Frontier.) From his

Admiral Adolphus Andrews. *Joseph Bilby.*

A "neutrality patrol" plane flies over a merchant ship in 1941. *Joseph Bilby.*

headquarters on the fourteenth floor of the Federal Building at 90 Church Street in Manhattan, Andrews had repeatedly petitioned his superiors in Washington for more patrol boats and weapons since the previous summer, to no avail.[73]

By December 1941, the United States, in cooperation with Britain, had been conducting "neutrality patrols" in the North Atlantic shipping lanes for some time, and several American ships—including the destroyer *Reuben James*, which had been torpedoed on October 31 while on convoy escort duty, inspiring an iconic folk song by Woody Guthrie—had been attacked and sunk by German submarines in the previous several months. Although a shipbuilding program had vastly increased the size of the American navy, the December 7 Japanese attack on Pearl Harbor, followed by the German declaration of war on the United States four days later, found the navy stretched thin along the Atlantic coast, especially as naval resources were transferred to what was considered the more imminent danger in the Pacific.[74]

As of January 1, 1942, Andrews's maritime force available to operate against what the admiral correctly perceived as a forthcoming surge in German submarine operations was a motley twenty-ship armada of Coast Guard cutters, gunboats and small "Yard Patrol" boats. His air resources were a few navy blimps and army air corps fixed-wing observation planes rather than combat aircraft, which only flew during the daytime, although some bombers were on the way. On December 22, Andrews notified his superior, Admiral Ernest J. King, that "should enemy submarines operate off this coast, this command has no forces available to take adequate action against them, either offensive or defensive." King, whose interests were more

Admiral Ernest J. King. *Joseph Bilby.*

focused on the Pacific theater of operations, had been appointed in 1940 to the newly created title of commander in chief of the United States fleet by President Franklin D. Roosevelt at the behest of former navy secretary Charles Edison, now the governor of New Jersey. He was unable to fulfill Andrews's request for reinforcements and was still trying to organize his own office staff in Washington.[75]

The irascible Admiral King's seeming lack of interest in coastal security and disregard of British advice—he thought them "haughty and arrogant"—in the opening months of the war has been attributed to his Anglophobia, apparently dating from his World War I service. King also believed (erroneously as it turned out) that coastal convoys defended by a mixed small-ship force was an ineffective tactic and that protection for overseas convoys with the limited number of destroyers at hand was more important than it was for merchant ships traveling along the coast. He did, however, allocate a few obsolete World War I–era destroyers to temporary rotating patrol duty to buttress the small "Yard Patrol" craft Andrews had on hand. The British found King's obstinacy mind-

A PT boat on "Yard Patrol" antisubmarine duty out of Cape May, New Jersey. *Joseph Bilby.*

A PT boat and crew on patrol out of Cape May, New Jersey. *Joseph Bilby.*

boggling, with one official declaring that he "found it extremely difficult to be polite about it."[76]

The German *Kriegsmarine* submarine force, led by Admiral Karl Doenitz, had been at war for over two years and was better organized for the task at hand, which Doenitz took as sinking as many Allied ships as possible to damage the resupply and reinforcement of Great Britain. He believed that an undersea assault on the United States' East Coast shipping would play an important role in this effort, especially in the sinking of oil tankers, and initiated Operation *Paukenschlag*, translated as "Roll of the Kettledrums," although it has also been called "Operation Drumbeat." The first wave of U-boats, *U-66, U-109, U-123, U-125* and *U-130*, departed from bases in France on December 16, 1941. British intelligence, which had cracked the German code, advised the Americans that the U-boats were on their way, and by the end of the month, sightings of periscopes and submarines traveling on the surface were reported off Canada and the coast of New England. Whenever possible, submarines traveled on the surface, where they could achieve greater speed than while submerged and also use less fuel.[77]

An Esso tanker off Sea Girt in 1939. This is a typical ship of the type the German U-Boats were out to sink. *Joseph Bilby.*

The first unmistakable sign that German submarines were closing in on New Jersey was the sinking of the tanker *Norness* on January 14 by *Korvettenkapitän* Reinhard Hardegan's *U-123*. The *Norness*, a Norwegian vessel sailing under Panamanian registry, was torpedoed shortly after midnight sixty miles off Montauk Point. The ship's captain, Harold Hansen, who survived along with thirty of his forty crewmen, commented afterward that

A German World War II U-boat. *Joseph Bilby.*

"no one was expecting a submarine so close in American waters." That "no one" apparently included the U.S. Navy, as the tanker's dying radio distress signal garnered no immediate response or reaction from shore.[78]

The patrolling blimp *K-3* discovered the *Norness* survivors floating in lifeboats twelve hours later and lowered coffee, cigarettes and sandwiches to them, in the first of 236 rescue missions performed by blimps during the coastal war, and then directed rescue ships to their aid. The following day, all available navy ships were detailed to escort convoy AT10 to Iceland and Northern Ireland, leaving East Coast shipping essentially defenseless. Admiral King, who advised coastal defenders to "make the best of what you have," claimed that a "curtain of steel" protected the convoy, and in that he was correct, as AT10 did not lose a single ship to German submarines. Unfortunately, the virtually undefended coast did not fare so well. One writer posits that King "did nothing" and bore much responsibility for the lack of reaction to the initial German offensive. Based on vague government assertions, the *New York Times* assured its alarmed readers, including those on coastal Long Island who were discovering wreckage and oil on their beaches, that the navy was equipped with "a powerful and numerous fleet of submarine chasers" that would halt further attacks.[79]

That fleet did not, of course, exist. After sinking *Norness*, Hardegan moved east toward New York City and New Jersey. On the way, he encountered the British tanker *Coimbra*, carrying a cargo of lubricating oil, one hundred miles due west of Sandy Hook and twenty-seven miles south of Southhampton, Long Island. As with *Norness*, bright lights on *Coimbra*'s starboard enabled *U-123* to easily target the vessel in an otherwise dense fog at 3:00 a.m. on January 15. Hit by a torpedo, the *Coimbra* exploded in a fireball of flaming oil. As a German observer noticed some crewmen running toward a deck gun on the ship's stern, *U-123* fired a second torpedo into *Coimbra*, breaking it in two and killing the captain and thirty-five crewmen. Ten men survived, six of them injured. The resulting fire was clearly seen from the south shore of Long Island by shore residents, who called the local police and Coast Guard stations. The survivors were rescued ten hours later. The *Coimbra* rests today in 165 feet of water and is still slowly leaking oil after all these years. After approaching the Ambrose channel and noting the Sandy Hook lighthouse, Hardegan headed south to his assigned hunting ground off Cape Hatteras, dodging a random bomber attack along the New Jersey coast on the way.[80]

The *Coimbra* slips beneath the waves off Long Island. *Joseph Bilby.*

The *Varanger* sinking off Atlantic City. *Joseph Bilby.*

And then war came to the New Jersey shore. On January 25, 1942, the *Varanger*, a Norwegian tanker, was sailing from Curacao to New York City with 12,750 tons of fuel oil aboard. Since Norway had been occupied by Nazi Germany, *Varanger* was operated, like *Norness*, "in allied service by the Norwegian Shipping and Trade mission, with offices in New York City." Shortly after 3:00 a.m., *Varanger*, sailing approximately twenty-eight miles from Brigantine Inlet, was hit amidships by a torpedo fired by *U-235* that knocked its radio room and a four-inch deck gun overboard. (Subsequent news reports located the ship anywhere from twelve to thirty-five miles from shore.) As lifeboats were being lowered, the tanker was hit by two more torpedoes in quick succession. Residents of Atlantic City reported that they were awakened by loud blasts in the middle of the night, and John W. Crothers of Sea Isle City told reporters afterward that "the explosions were so severe that they rattled his house and almost tossed him out of bed."[81]

Varanger split into three pieces, but all hands escaped in lifeboats. All forty crewmen were rescued by sixty-year-old fishing boat captain Dewey Conchetti of Sea Isle City. Conchetti, a fisherman for thirty-five years, had seen a "big flash" on the horizon as he left port early that morning in his

Fishing boat captain Dewey Conchetti of Sea Isle City, New Jersey (right), with crewman Edward Elisano (left) rescued the crew of the *Varanger. Joseph Bilby.*

Opposite, top: Varanger crewmen at the U.S. Immigration Service station in Gloucester City, New Jersey, where some were photographed as they were visited by the Norwegian Consul from Philadelphia, Mathias Moe (with glasses) and assistant U.S. immigration inspector Adolph Schiavo. *Joseph Bilby.*

Opposite, bottom: Varanger crewmen get a medical checkup at Gloucester City, New Jersey. *Joseph Bilby.*

boat, the *San Gennaro*, to fish for cod but did not think it important. He later came upon *Varanger*'s two lifeboats at sea and called on fellow fisherman Captain Dominick Constantino of the *Eileen* for assistance. The two fishing boats towed the oil-spattered tanker crew to the U.S. Coast Guard station at Sea Isle City, arriving there at 12:45 p.m. Once in the station, the *Varanger* crewmen received "kerosene baths" to remove oil from their bodies and were treated by local doctor Alexander Stuart for superficial injuries. Doctor Stuart recalled that the *Varanger* survivors were "a tough gang. They could take the ordeal of their sinking." Conchetti, who was proud of his part in the rescue, told inquiring reporters that his son was currently serving in the United States Army.[82]

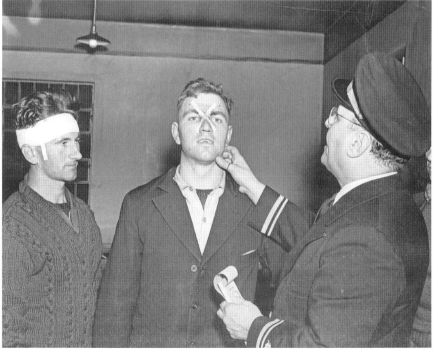

The survivors exchanged their oil-soaked clothes for garments donated by local citizens and the Red Cross and spent the night on cots in the basement of Saint Joseph's Catholic Church. One man wandered out the door but was returned by the Coast Guard, which did not want the crew talking to reporters until the official navy version of the incident was released. The following day, the lucky *Varanger* sailors were transferred across the state from the Coast Guard station to the U.S. Immigration Service station in Gloucester City, New Jersey, where some were photographed as they were visited by the Norwegian Consul from Philadelphia, Mathias Moe, and assistant U.S. immigration inspector Adolph Schiavo. The hulk of the *Varanger*, nicknamed the "twenty-eight-mile wreck," still sits off the New Jersey coast at a depth of 145 feet and is a popular dive site for advanced divers, as well as an excellent fishing location. The ship's bell was recovered by a diver in 1978.[83]

Varanger was just the beginning. Around 6:45 p.m. on February 5, 1942, the Socony-Vacuum oil company tanker *India Arrow*, bound from Corpus Christi to New York with a cargo of 88,369 barrels of diesel fuel aboard, was torpedoed by *U-103* about twenty miles southeast of Cape May. The ship caught fire immediately, and as the crew scrambled for the lifeboats, *U-103* surfaced and began to fire its deck gun at the sinking vessel at a range of 250 yards, hitting it seven times. The radio operator had managed to send a message that the ship had been torpedoed, but not its location. The Eastern Sea Frontier headquarters tried to figure out where the *India Arrow* was located and dispatched a patrol boat. Attempts to have air resources search for survivors were largely bungled by differences of opinion on areas of responsibility and the failure to coordinate between the navy and the army air corps unit stationed at Langley Field.[84]

Only two lifeboats were successfully launched from the *India Arrow*, and one of those was struck by debris resulting from deck gun shellfire and sank with all on board. Spilled oil ignited as well, and the sea was soon afire. Captain Carl S. Johnson, who was on the bridge when it collapsed, jumped, hit the water and was "miraculously" swept into the only remaining lifeboat by a wave. Johnson's boat managed to escape the burning sea and pick up another nine survivors, including Edward J. Proehl of Jersey City, and then, with oars and a sail, set out for the New Jersey coast. Hours later, the bedraggled survivors saw the lights of Atlantic City beckoning in the distance. They flashed their flashlights at passing ships but were ignored, as the fear of submarines falsely signaling caused most captains to ignore lights in the dark.[85]

At 6:30 a.m. on February 6, the survivors of the *India Arrow* were finally rescued by Frank D. Marshall's fishing boat *Gitana*. Marshall and his mate

The *India Arrow. Joseph Bilby.*

John Shaw were fishing about twenty miles southeast of Atlantic City when they spotted the lifeboat in the distance and sailed toward it. Marshall recalled that the men he saved were a "sorry looking crew" soaked with oil and water and mostly barefoot. He brought them into the Atlantic City Coast Guard Station, where they were cleaned up, fed, issued new clothes and received medical attention. The twenty-six missing men from the *India Arrow* crew, including Jerseymen Nicholas Hetz of Camden and Carl Hunnergarth of Hoboken, were never found.[86]

The USS *Jacob Jones*—named after a career naval officer, Delaware native, Barbary pirate and War of 1812 hero—was a "*Wickes* Class" destroyer commissioned in 1919. It replaced a previous *Jacob Jones*, the destroyer noted in Chapter 2 as sunk by a German U-boat off Queenstown, Ireland, on December 17, 1917, during World War I. Fifty of these older *Wickes* Class destroyers were traded to Great Britain in exchange for naval base use as part of the lend-lease program enacted in March 1941 and were mostly used as convoy escort vessels, while others, including the *Jacob Jones*, were retained and deployed for coastal defense.[87]

Assigned to the Eastern Sea Frontier for temporary duty in early 1942, the *Jones*, nicknamed by its crew the "Jakie," was detailed by Admiral Andrews, desperate to take some defensive measures, to roving anti-submarine warfare patrol duty, a tactic that had failed during World War I, when President Woodrow Wilson, not particularly known for his military acumen, accurately described it as "hunting the hornets all over the farm." On February 22, while patrolling near Ambrose Light, the destroyer's commander, Lieutenant Commander Hugh P. Black, dropped forty-seven depth charges on what he believed to be an enemy submarine during a five-hour period but without any tangible result.[88]

On February 27, the *Jones* left New York and steamed south along the New Jersey coast, initially assigned to patrol an area between Barnegat and Five Fathom Bank Light Buoy. Black was ordered to sail up to forty-five miles out at sea during the day and "five miles outside the lighted buoy line" at night. While underway, Lieutenant Commander Black received orders to extend his patrol area south to Cape May and Delaware Bay. At 3:30 p.m. that afternoon, the destroyer came upon the burning wreck of the *R.P. Resor*, torpedoed during the previous night. The *Jones* circled the wreck of the *Resor* for two hours looking for survivors, while crewmen lined the deck, staring at the burning hulk, but found none and then continued south. As darkness fell, Black turned off his running and navigation lights and continued south.[89]

Around 5:00 a.m. on February 28, *U-578*, the same submarine that sank the *R.P. Resor*, fired two torpedoes at the *Jones*, then sailing around thirty miles off Cape May on a bright moonlit night, and they hit the destroyer's port side almost simultaneously. Chaos ensued when the destroyer's ammunition magazine exploded as the first torpedo hit. Apprentice seaman Adolph Storm recalled that "there was a tremendous concussion and a flash of flame and I was knocked down." Seaman Joseph Tidwell, who was in the galley at the time, remembered, "Pots and pans rained down upon our heads." When Seaman Storm reached the deck, he found confused sailors trying to launch lifeboats, yelling, "What happened?" over and over. The *Jones* was split into three pieces.[90]

Only about twenty-five crewmen survived the initial blasts and took to life rafts, as the ship's lifeboats were snagged in the wreckage and unable to launch. The crewmen who managed to escape carried the sole surviving officer, who was so badly wounded that he was "dazed and practically incoherent," off with them. (He did not survive.) Many survivors of the initial explosions were killed when the *Jones*'s depth charges exploded underneath them as the ship went down, creating a one-hundred-foot-high column of water. Shortly after 8:00 a.m., an army observation plane piloted by First Lieutenant R.L. Blackburn spotted several life rafts bobbing around in the ocean and alerted the Coast Guard patrol boat *Eagle 56*, one of Henry Ford's surviving World War I project ships, from Cape May to look for survivors from the 120-man crew. By 11:00 a.m., the boat, struggling against high seas, had picked

Newspapers across the country informed the public of the fate of the *Jacob Jones*, torpedoed off Cape May, New Jersey. *Joseph Bilby.*

up twelve survivors, including seventeen-year-old Apprentice Seaman John Struthers of Trenton. One of them, Seaman Carl Smith, died on the way to shore. Two more days of searching by aircraft and ship failed to find either survivors or bodies. The *Jacob L. Jones* was the first navy warship sunk on the Eastern Sea Frontier. A postwar review of the tactics that dispatched the destroyer on its mission concluded that "a roving patrol was worse than useless; it resulted only in the loss of *Jacob Jones*."[91]

In the early morning of March 9, 1942, the *Cayru*, a Brazilian merchant ship bound for New York with a crew of sixty-six and carrying a cargo of leather, oil, cotton and cacao, as well as fourteen passengers, was sailing up the New Jersey coast. The ship had its lights blacked out, as German submarines had sunk several other Brazilian ships, disregarding the country's neutrality, in the recent past, but it was not zigzagging. Weather conditions were gloomy, described as "no moon, dark and cloudy, poor visibility," when, at around 2:30 a.m., *Cayru* was hit by two torpedoes fired by *U-94* in rapid succession. The first failed to detonate, but the second exploded amidships and broke the vessel in half and it began to founder.[92]

The passengers and crew quickly abandoned ship in four lifeboats, and then *U-94* surfaced amid them. An officer appeared in the conning tower and, "in broken English with an unmistakable German accent," asked the

Cayru crew survivors in New York. *Joseph Bilby.*

survivors the name of the ship and its destination. When assured there were no passengers and crew still on board, he fired another torpedo into *Cayru*, which rapidly sank.[93]

One *Cayru* lifeboat with twenty-two crewmen and four passengers was picked up later in the day by a Norwegian freighter, which brought the survivors into New York. They included Mrs. Willie Saunders De Souza, a Virginia woman who had married a Brazilian, and her fifteen-year-old daughter June, who were subsequently interviewed by the press at the Hotel Wellington in Manhattan. Otto Jaegers, a World War I veteran who was also in the boat, recalled that "the bravest of the lot was the 15-year-old girl, June De Sousa [*sic*]. She never let a peep out of her, nor complained from the time we were left in the lifeboat until the time we were rescued."[94]

Another *Cayru* lifeboat, with six survivors and one body, was picked up by an American navy minesweeper on March 11 and brought into New London, Connecticut. All those aboard had frozen feet and "were suffering greatly from exposure." Two were able to walk with help from sailors, but the others were carried ashore on stretchers. Fourteen other occupants of that boat had been

Cayru survivors Mrs. Willie Saunders De Souza and her fifteen-year-old daughter, June, being interviewed by the press at the Hotel Wellington in Manhattan. *Joseph Bilby.*

swept overboard in bad weather. No other survivors were ever found, nor has, to this date, the location of the *Cayru*'s hulk been identified.[95]

The sinking of the *Cayru*, the fourth Brazilian ship sent to the bottom by German submarines within a month, sparked riots in Rio de Janeiro, as

A lifeboat from the *Cayru*. It was picked up with six survivors and one body by an American navy minesweeper on March 11, 1942, and brought into New London, Connecticut. *Joseph Bilby.*

thousands of Brazilians demonstrated against German nationals and businesses and cried "down with Germany, viva Brazil." A man who had the temerity to announce that he was "honored to be a German" was "attacked and brutally beaten." The Brazilian government seized German bank accounts and property, initiated sanctions against the Axis countries and declared war on Germany in August.[96]

As the *Cayru* slipped under the waves, the *Gulftrade*, an oil tanker heading to New York from Port Arthur, Texas, was sailing much closer to the New

The *Gulftrade* sinking. *Joseph Bilby.*

Jersey coast than the Brazilian vessel, when Captain Torger Olsen decided to turn on his ship's navigation lights to avoid a collision as *Gulftrade* approached the busy sea lanes south of New York City. He recalled that he "saw we were up to Barnegat and I thought they shouldn't be able to get us now anymore." As the tanker passed Barnegat Inlet around three miles out at sea just after midnight on March 10, it was hit amidships by a torpedo fired by *U-588*. Olsen was in his quarters looking at a chart at the time, and his room collapsed around him in the explosion. When the captain reached the deck he saw "flames…ninety-six feet high." Those flames, which were fortunately quenched shortly afterward as a massive wave crested over the ship, were clearly visible from shore to any local residents still awake at that hour.[97]

The *Gulftrade* split in half. Olsen and six of his men got off in a lifeboat and saw other crewmen launch two lifeboats from the bow section of the ship. Those two lifeboats and their occupants were never seen again. One group of crewmen decided to stay on the stern section as the bow drifted away. Chief engineer Guy F. Chadwick recalled that they "huddled behind barrels, in fear that the submarine would come to the surface and machine-gun or shell them." Olsen's party was picked up by the Coast Guard cutter *Antietam*, which was cruising nearby, after half an hour, but those who remained on the drifting stern were not rescued for another three hours, when the heroic crew of the naval patrol boat *Larch* risked their own lives to get them safely off the wreck. All of the survivors arrived at Pier 6 in New York City the following day. The bow section of *Gulftrade* drifted three miles to the east and grounded, where it was visible from shore. An air patrol spotted an empty lifeboat bobbing in the sea three

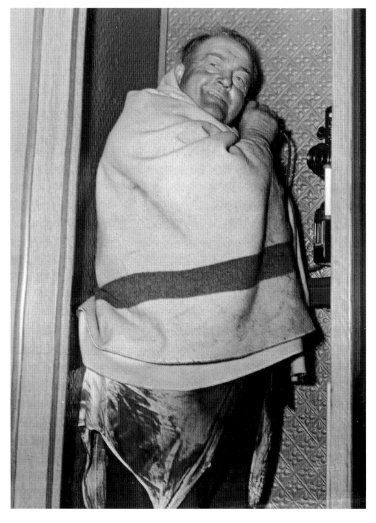

Edward Johnson, second mate on the *Gulftrade*, telephones his family to tell them he is safe in New York City. *Joseph Bilby.*

Captain Torger Olsen of the *Gulftrade* (right) survived, only to die on October 21, 1943, when his next ship, the *Gulfland*, was in a collision with another tanker, the *Gulfbelle*. *Joseph Bilby.*

miles north of the stern. In the end, sixteen of the crew survived and eighteen were lost. Ironically, Captain Olsen survived the attack only to die on October 21, 1943, when his next ship, the *Gulfland*, was in a collision with another tanker, the *Gulfbelle*.[98]

On March 13, the Chilean freighter *Tolten* was steaming toward New York to pick up a cargo after dropping off a load of nitrates in Baltimore. Chile was a neutral nation, and there was some controversy over whether the country's ships should obey American navy orders to sail with lights out.

The Chilean government, sympathetic to Germany, had ordered its vessels to keep their lights on and visibly fly Chilean flags to avoid being torpedoed but apparently advised the *Tolten*'s captain, Aquiles Ramirez, to comply with American instructions, although there is evidence he initially did not, until ordered to do so by a navy patrol boat.[99]

In the early morning hours of March 13, *Tolten* was torpedoed by *U-332* sixteen miles off Barnegat Inlet and reportedly sank within six minutes. There was only one survivor of the twenty-seven-man crew, electrician Julio Faust, who was spotted on a life raft by a patrol plane from Langley Field. The Coast Guard cutter *Antietam* and the navy patrol boat *Larch*, both of which had deployed for the *Gulftrade* rescue, were dispatched to the scene, while several blimps were dispatched from Lakewood. Around noon, blimp *L-12* reported seeing "two bodies 17 miles east of Sea Side Park" and then another body, and finally, a "life raft with a live man [Julio] (Faust) on it." *Antietam* picked up Faust, who was "suffering from exposure, shock and bruises," along with five unidentified floating bodies, and transferred them to *Larch*. In the aftermath, navy secretary Frank Knox imposed an informational media gag rule on future sinkings, adopting the Canadian policy of not providing the names and locations of the ships sunk but merely their national identity and vague definitions of their size to the media.[100]

The sinking of the *Tolten* echoed far beyond the New Jersey coast. As in Brazil following the *Cayru* sinking, it gave rise to demonstrations, although the government tried to suppress them. In Santiago, crowds of young people "smashed windows of the pro-Axis newspaper Chileo, a German bar, the Casa Hombo, a Japanese novelty shop and the Casa Oberpauer, a German clothing store," while "truckloads of youths drove through the streets shouting "Down with the Axis.""[101]

In contrast to Brazil, the Chilean government downplayed the incident, keeping information as much out of the papers as possible and, when pressed, blaming the United States' orders to sail blacked out as the cause of the sinking. A Chilean admiral stated that he was confident there was "no deliberate intent" to sink the *Tolten*, as the "submarine could not know the flag under which the *Tolten* was sailing." Chile eventually broke off relations with the Axis powers but never declared war on Germany. With the conflict in its last stages in 1945, the country declared war on Japan.[102]

Although detailed news stories of German submarine attacks along the American coast dried up overnight following Knox's directive, the carnage did not end. Admiral Andrews was reluctant to order coastal convoys, as he claimed that he did not have the escort and support vessels necessary, and so ships continued to travel alone. In the early hours of March 14, Captain Gardner F. Clark, skipper of the *Lemuel Burrows*, which was steaming north on a route to Boston, ran his ship very close to shore, later claiming he could clearly see the lights of Atlantic City, as it was thought that the water there was too shallow for submarine activity. Clark was wrong. Although he was sailing with all lights out, at around 2:00 a.m., members of the army air corps' 104th Observation Squadron, based in Atlantic City, heard "heavy explosions" as the *Lemuel Burrows* was torpedoed about twelve miles off Absecon Inlet. While Clark and his crew abandoned ship, mostly in life rafts, as several lifeboats had been destroyed in the attack, the vessel was hit by two more torpedoes, creating a wave that swamped a life raft. Six hours later, the steamer *Sewall's Point* picked up six survivors clinging to the overturned lifeboat about five miles from Atlantic City and shortly afterward saved another eight men drifting in life rafts. Twenty crewmen died. The following day, a patrol boat towed in an overturned lifeboat to the Hereford Inlet Coast Guard Station and discovered a body jammed in it. A few days later, yet another body was discovered on the Wildwood beach, and part of the deck of an unnamed ship with three bodies on it floated into Wildwood. On March 20, a naval diving team descended onto the wreck hoping to find the ship's papers. They were unsuccessful.[103]

The *Lemuel Burrows*'s second engineer blamed the sinking on the lights of Atlantic City, which he said silhouetted his ship against the horizon, claiming, "It was like Coney Island. It was lit up like daylight all along the beach." It should be noted, however, that Captain Clark deliberately ran his ship close to shore and that silhouetting from the horizon glow of shore lights was generally estimated to extend ten miles, although some accounts suggest as much as twenty-five miles, out to sea, depending on weather conditions and the intensity of the shore lights. An executive order from President Roosevelt on February 19 gave the military authority to "assume control over all lighting on the seacoast" as a defensive measure against enemy submarines. On March 4, that supervision of coastal lighting was declared a navy responsibility, and several days later, Admiral King sent a message to Admiral Andrews asking him to "take such steps as may be in his province" to control the lighting levels on the seacoast. On March 14, King actually restricted Andrews's authority to deal with the matter by telling him that blackouts were "not considered necessary" and that a "dim-out," shutting down the brightest lights, would suffice. To further confuse the issue, the army's II Corps, also responsible for the defense of the New York–New Jersey region, issued its own dim-out order. Admiral King, in contrast to leadership in Britain and Germany, never ordered a blackout along the coast. The dim-out was in effect when the *Lemuel Burrows* went down.[104]

On April 1, the navy's public relations office issued a morale-boosting statement declaring that coastal defense efforts had resulted in twenty-eight U-boats "sunk and presumed sunk" between the Canadian border and the Gulf of Mexico. Four of these submarine kills were credited to the army air corps and twenty-four to the navy. In reality, not a single U-boat had been sunk. That month, Admiral Andrews ordered ships moving along the coast in the Eastern Sea Frontier to only travel by day, a direction that was not universally followed.[105]

On April 28, the *Arundo*—a freighter sailing out of New York to North Africa with a stop at Capetown with a cargo of military supplies, including 123 trucks and jeeps, two locomotives and five thousand cases of beer—had a navy armed guard detachment manning a four-inch deck gun, two twenty-millimeter antiaircraft guns and several machine guns aboard. When the ship was about sixteen miles off Manasquan Inlet, it was hit on its starboard side by two torpedoes fired by *U-136*. Six crewmen were killed, four of them by one of the locomotives sliding off the deck. The hulk of *Arundo* lies in an area known as Wreck Valley, where, interestingly, it is the only victim of a torpedo attack. Other ships that sank in the area during World War II were the victims of collisions. A popular site for scuba divers, the *Arundo* wreck, littered with jeep tires and beer bottles, claimed the life of a female diver in a 2007 accident.[106]

On April 30, the *Bidevind*, a Norwegian freighter, was heading toward New York from India with a cargo of "wool, goat skins, ginger, fragrant oils, seeds, gum and nuts." When *Bidevind* was about fifty miles at sea off Manasquan Inlet, just before 11:00 p.m., it was struck by a torpedo fired by *U-752*. As the ship began to list, it was hit by another torpedo shortly afterward and sank after fifteen minutes. Navy divers found the wreck of the *Bidevind* on September 30, 1943, reporting that it was in 190 feet of water. It has since largely disintegrated.[107]

Admiral King, who had initially been against the use of coastal convoys, believing them ineffective, began to change his mind. By May, Admiral

Opposite, bottom: K-Class blimps take off on a mission from the naval air station at Lakehurst, New Jersey, for coastal patrol in World War II. The forty-foot-long K-Class airships were equipped with communications equipment and instruments facilitating night flying, as well as radar and other detection devices. They were armed with four depth charges and a .50-caliber machine gun and had a crew of ten men. *Joseph Bilby.*

An Atlantic City party fishing boat in 1940. Civilian vessels like these, as well as yachts, were manned by volunteer submarine spotters off the New Jersey coast in 1942. *Joseph Bilby.*

Andrews had access to a variety of escort vessels, including a few destroyers on temporary duty, Coast Guard cutters, gunboats, anti-submarine trawlers from Britain and converted private small craft, as well as an increasing number of aircraft, including navy blimps, some B-25 bombers and Civil Air Patrol unarmed observers, to conduct and protect small coastal convoys, a much safer way to travel than alone with submarines in the area. Assisting this force in submarine detection were a number of private yachts and fishing boats attached to the Coast Guard and called the "corsair fleet." These "bucket brigade" convoys used inland waterways when possible and only traveled by day.[108]

On May 25, *Persephone*, a Panamanian tanker, was on its way from Aruba to New York with a load of ninety thousand barrels of fuel oil in a convoy escorted by navy and Coast Guard ships and aircraft. Unfortunately, *Persephone* was the last vessel in the convoy, which traveled close to shore, as the shallow water was thought to be a deterrent to submarine activities. *Persephone* was only about three miles at sea from Barnegat Light at 3:30 p.m. when it was hit in the stern by two torpedoes fired from *U-593*, and the resultant explosion was distinctly heard on shore. *Persephone*'s stern began to sink as the crew abandoned ship, and nine of the thirty-seven men on board went down with it. The bow remained above water, however, enabling Captain Helge Quistgaard to save his navigation equipment and, after he was picked up by a Coast Guard vessel, return to the ship to rescue twenty-three bags of United States mail.[109]

People ashore watched the unfolding drama, and "the principal of the Seaside Park Grade School and the teacher of the one-room Barnegat City School led the curious students to the beach, where they watched the ship sink while a Navy blimp out of Lakehurst Naval Air Station searched in vain for the submarine." Local first aid squad members rushed to the Barnegat City Coast Guard Station to aid the injured and wash oil off the surviving crewmen. Two members of the crew, a Swede and a Dane, decided to make Barnegat their new home, but the bad luck of the *Persephone* followed them and one died of a heart attack the following year and the other in an automobile accident in 1945.[110]

Persephone's bow remained above water and was towed to New York, where twenty thousand barrels of the ship's fuel oil cargo were salvaged. The bow was eventually towed to Baltimore, where it was combined with the stern of another damaged ship to create a new vessel. The stern remains off Barnegat to this day, a "scattered junk heap sitting in 55 feet of water," where it is used

A depth charge, the most effective anti-submarine weapon, explodes in the Atlantic Ocean. *Joseph Bilby.*

by party fishing boats seeking blackfish and sea bass, as well as scuba divers hunting lobsters.[111]

Since *Persephone* was traveling in an escorted convoy, *U-593* was immediately pursued and attacked by escort vessels dropping depth charges and covering aircraft dropping bombs. The German submarine escaped but was damaged in the attack and had to limp back to France for repairs. Things were beginning to change, and the New Jersey shore—and, by implication, the entire East Coast—was no longer the "American Shooting Gallery" that it had once been for prowling U-boats.[112]

On June 22, the Argentine freighter *Rio Tercero* was steaming south eighty miles offshore from Brigantine with a varied cargo of rayon cloth, asbestos fiber, insulation, newsprint and clocks bound for Buenos Aires. It was not traveling in a convoy, as Argentina was a neutral country and, with a significant number of people of German and Italian extraction, actually tilted in sympathy somewhat toward the Axis powers, and its distance from the coast certainly protected it from being a silhouetted target. Just to be on the safe side, however, the *Rio Tercero*, perhaps considering the fate of the *Tolten* several months earlier, as well as the Argentine ship *Victoria*, sunk in April, was emblazoned with thirteen Argentine flags along its hull to reinforce its neutrality to any lurking German submarine.[113]

Unfortunately for the crew of the *Rio Tercero*, *Kapitanlieutenant* Hans Heinz Linder, the commander of *U-202*, either ignored or could not see any indications of the ship's national origin. Linder, whose submarine had left Brest on May 27 and had recently landed four German saboteurs on Long Island (a mission that Linder referred to as his "special task" and would have subsequent repercussions in New Jersey), attacked the Argentine vessel in broad daylight. The German commander noticed that the steamer was not zigzagging and later claimed that he noticed no flag or "special markings" on the hull before he fired. The torpedo slammed into the side of the ship, killing four crewmen instantly and destroying the starboard lifeboats. Within ten minutes, the *Rio Tercero* went to the bottom, taking another unlucky sailor with it. The rest of the crew, including Captain Luis Pedro Scalese, managed to escape in the remaining lifeboats.[114]

Linder surfaced his submarine and sped toward the bobbing lifeboats and, when he reached the site of the sinking, noticed "a great number" of bales of newsprint cargo floating on the surface. He engaged Scalese in a rather stilted conversation, due to language difficulties, and, reminiscent of World War I operations off New Jersey, ordered the Argentine captain to come aboard *U-202*. The German officer asked Scalese questions as to his ship's name and cargo and requested the ship's log and then treated him to "three glasses of cognac and a pair of shoes" before releasing him back to his lifeboat. Linder concluded, erroneously as it turned out, that the *Rio Tercero* was, although Argentine, "in enemy service." Scalese and his surviving crewmen were eventually picked up by the U.S. Navy and brought into New York. After dodging an American bomber attack, Linder headed south to Cape Hatteras.[115]

The regulation on not disclosing the full stories of torpedoed ships was discarded by the media in the case of the *Rio Tercero*, as the sinking created international news and the Argentine foreign minister officially announced it. Captain Scalese publicly proclaimed his gratitude to the American military personnel who saved him and his surviving crewmen, and the Argentine ambassador to the United States, Felipe A. Espil, released a statement expressing his government's thanks for "the prompt and decisive cooperation of the naval and air forces of the United States in the task of saving the victims, almost all of them of Argentine nationality."[116]

Argentine diplomats in Germany delivered a protest note to the German government in Berlin and got a rapid reply, indicative of "Germany's desire

Captain Luis Pedro Scalese of the *Rio Tercero* speaks to the press in New York City. *Joseph Bilby.*

to reach a quick settlement of the incident." The Germans claimed that the "error that arose because the commander of the U-boat and his officers had been unable to recognize the Argentine flag on the *Rio Tercero*" was responsible for the attack. They agreed to pay compensation, as they had with the *Victoria*, but refused to "salute the Argentine flag" as part of the settlement, as that was not part of the policies of the "New Germany."[117]

In the aftermath, there were "small anti-Axis incidents...in the streets of Buenos Aires...Some windows were broken in an Italian bank." A group advocating continued neutrality was attacked and a German-owned electrical supply company was stoned, but Argentine neutrality continued, albeit amid domestic political turmoil. Under American pressure, Argentina finally broke diplomatic relations with Germany on January 26, 1944, and declared war on Germany on March 27, 1945.[118]

As summer waned along the New Jersey coast in 1942, so did U-boat activity, in the face of increased and more effective defensive measures, with effective convoys, more small boats and aircraft on patrol and better radar and sonar detection. A German naval officer recalled after the war, "In summer of 1942, the gradually increasing defense measures of the United States in the coastal waters of the Western Atlantic had forced the U-boats to shift their main hunting grounds to the mid-Atlantic." On July 19, Admiral Doenitz, realizing that the tide had turned against him, withdrew his last two U-boats, which were hunting off Cape Hatteras, from the Atlantic coast. Eight days later, he officially changed his U-boat main mission back to attacking transatlantic convoys. Sinkings along the entire length of the Eastern Sea Frontier dropped from thirteen in June to three in July and none for the rest of the year. As Theodore Taylor noted in his 1958 book on the coastal war, "the blitz was over."[119]

Although on June 23 the tugboat *John R. Williams* hit a mine that had been sowed by *U-573* off Cape May several weeks before and sank immediately, losing fourteen of its crew of eighteen men, the war off the Jersey shore effectively ended with the sinking of *Rio Tercero*. As bad as it was, it had never been as intense as the submarine warfare farther south, particularly off Cape Hatteras and in the Gulf of Mexico and the Caribbean, where the submarine threat remained serious for a while after it abated along the North American coast, but it certainly brought war home dramatically to New Jerseyans, including former U.S. Navy secretary Governor Charles Edison and his war cabinet, which met weekly in Trenton to discuss and initiate implementation of home front defenses, and found themselves both cooperating and conflicting with the military.[120]

Chapter 4

THE JERSEY SHORE COPES

"SOME AMERICAN AIRCRAFT POSSIBLY USED THE WHALE AS A TARGET"

O n Friday, January 30, 1942, there were plenty of diversions to be found at the Jersey Shore. Those in search of a respite from the incessant drumbeat of war news could swing to the big band tunes of Harry James at the Sea Girt Inn. Less adventurous types could thrill to the plight of Walter Pidgeon and Maureen O'Hara in the local movie premiere of *How Green Was My Valley* at the luxurious Mayfair Theater in Asbury Park. Even beleaguered housewives who were too immersed in their domestic duties to enjoy a night on the town could plan ahead for a four-day "defense cooking course" being held the following month at Asbury Park's Convention Hall.[121]

No amount of escapism, however, could blot out the ominous headline that bannered the front page of the *Asbury Park Press*, the Jersey Shore newspaper of record: "New Wave of U-Boat Attacks Seen Along the Atlantic Coast." The news account warned of a likely new wave of submarine activity, despite the alleged increasing effectiveness of American countermeasures. Unlike in 1918, this time around, people were not rushing to the Asbury Park boardwalk because "the possibility of a Hun raid appearing off the coast has aroused the keenest interest." The times were much more sober.[122]

New Jersey residents were already jittery about submarines, following a series of attacks that month off the East Coast, most recently against the tanker *Varanger* close to home earlier that week. Building on the public's concerns, authorities quoted in the January 30 news article noted that despite the lull in activity since the *Varanger* attack, enemy submarines continued to

New Jersey governor Charles Edison, son of the inventor Thomas Edison. *Joseph Bilby.*

operate along the East Coast as far south as Florida. U-boats fresh from Germany would now be replacing those that made the first attacks on coastal shipping during the week of January 12, officials warned. In fact, there was speculation that "the Germans might try to keep more than a score of U-boats off the Atlantic coast in an attempt to force the Navy to divert warships from the North Atlantic convoy route to meet the new menace." Indeed, that was one of the German goals, along with damaging American morale and interrupting oil supplies.[123]

These and other potential problems weighed heavily on the state's political leadership as well and were consistent topics of conversation in the weekly War Cabinet meeting convened by Governor Charles Edison, son of the famed New Jersey–based inventor. Edison even went so far as to build a concrete bombproof bunker at the Edison West Orange laboratory to protect his father's papers from possible air attack. It still exists.

For New Jersey shore residents in particular, the Nazi U-boat attacks posed a palpable danger that might result in stringent blackouts, overall civilian anxiety and thus a serious threat to tourism, the lifeblood for coastal communities almost totally dependent on summer business and just beginning to recover from the Great Depression. The U-boat menace had genuine urgency, "for in early 1942 New Jersey came as close as it ever did to front-line status."[124]

Local residents Nat Brooks and Terry Hughes recalled the chaos that ensued after a successful U-boat strike: "Fire whistles sounded often during the night. Local fishermen raced out to try to rescue seamen from the ocean on fire. Rumor flew [that] Germans had landed on the once pristine beach, now a goopy mess resulting from the torpedoed oiler."[125]

Concerns about the U-boat menace were part of a broader, gnawing uncertainty about the war. During those early months of 1942, the United States appeared to many, including military and civilian leaders, to be distressingly vulnerable to Axis attack, and the picture only became grimmer after a February 14 meeting of Allied leaders who convened to discuss their concerns that three powerful Nazi fighting ships, once thought imprisoned in a French port, were now free to challenge the Allies in the Atlantic.

The three Nazi vessels—the battleships *Scharnhorst* and *Gneisenau* and the cruiser *Prinz Eugen*—dashed from their often-bombed refuge in Brest, brushed aside British surface and air interception and sped for the Nazis' Helgoland Bight, according to news reports. The ships posed a potentially grave menace to the sea lanes from America to England and Russia and could provide a deadly complement to U-boat attacks.

Aside from the threat to sea lanes, some officials saw a new and direct hazard to American cities. New York's mayor, Fiorello LaGuardia, told reporters the war had come closer to home as a result of the escape of the Nazi sea squadron and that "incendiary bombings from the air were possible." LaGuardia's warning came shortly after Senator Scott W. Lucas (D-Ill) asserted that if Japan won the ongoing struggle in the Pacific "she may turn to Alaska and drop poison gas bombs on American mainland cities."[126]

President Franklin D. Roosevelt himself only heightened concerns at a February 17 press conference when the former assistant secretary of the navy addressed the possibility that German submarines might be operating from a hidden base in the Caribbean, evoking the comments of his World War I boss, Josephus Daniels. "President Warns Country's Shores Can Be Attacked," trumpeted the *Asbury Park Press* headline. The president went so far as to warn that the enemy "could shell New York or drop bombs on Detroit under the right circumstances." If the president believed that New York could be shelled, that was not good news for New Jersey shore towns.[127]

The nightmarish scenario reflected how quickly life on the homefront had changed since the bombing of Pearl Harbor only two months before, and few would have predicted the traumatic turn of events. In the years prior to U.S. involvement, the overwhelming consensus in New Jersey was to avoid participation in another war, as the aftermath of World War I had left a bad aftertaste for a seeming majority of Americans. During the 1930s, the state's congressional delegation voted for a series of neutrality acts that severely restricted the sale of arms to belligerents, and students at many New Jersey colleges held strikes for peace. Reflecting a schism throughout the country, the state's residents were divided over Roosevelt's efforts first to support

Great Britain and then the Soviet Union, after that country was invaded by Hitler in June 1941.[128]

The war "over there" became the subject of heated discussion in many New Jersey homes. Dorothea Smith, a summer resident of Harvey Cedars, remembered such a debate with her children, Dick, Dorothea and Steve. "I don't see why we have to get into war," Mrs. Smith said during Sunday dinner. "Here in this country we have Italians, French, Germans, English, Slavs, Russians—all living together peaceably. Because they can't get along together in Europe is no reason for us to get mixed up in that mess." Her son Steve countered, "Now that Hitler is in, what do you think we ought to do—just let him run the world, conquer and kill innocent people?"[129]

By the fall of 1941, New Jersey, like the rest of the country, existed in an eerie state of stasis, and citizens tried to distract themselves from the ominous news of German and Japanese aggression. Along the Asbury Park boardwalk during that unsettling autumn, Frank Sinatra crooned his melancholy hit "This Love of Mine" from jukeboxes and radios. Sinatra's "yearning cries" captured the popular mood of apprehension and a longing for security.[130]

The unsettling lull had ended abruptly on Sunday, December 7, 1941. Many New Jerseyans were listening to the Giants football game on WHN when news of the Japanese attack on Pearl Harbor interrupted the broadcast. Across the state, frantic listeners rushed to their telephones to commiserate with loved ones, while thousands of men mobbed recruiting stations the next morning.

One diner customer reflected the popular public sentiment: "The Japs started this war and we'll give them plenty." Underneath the bravado and flurry of volunteer activities, however, lurked a mounting sense of fear and misgiving. The news in the days after the attack was grim, including reports of the sinking of two British warships and the Japanese invasion of Luzon on Tuesday, the U.S. declaration of war against Germany and Italy on Thursday and the Japanese attack of Guam and Midway on Friday.[131]

Although it seems unlikely in retrospect, New Jerseyans took seriously the warnings of direct Axis attacks on the state: "There was a pervasive belief that New Jersey might be attacked from the air, just as had happened in Hawaii." During the first week following Pearl Harbor, there were numerous false alarms; a rumor spread through Newark that an attacking force of airplanes was headed toward the city, and police headquarters received "3,000 calls from panicked residents trying to find out where to find air raid shelters." At Newark's Bamberger's Department Store, civil defense uniforms were quickly piled on shelves, while Kresge's Department Store reported "brisk sales of blackout coverings for windows."[132]

Paranoia about an Axis attack manifested itself in various ways at the Jersey Shore. Panic struck the communities between Belmar and Point Pleasant early in the morning of February 16, 1942, when a strong odor of oil awoke hundreds of residents, leading them to believe a tanker had been lost offshore. The odor, which was strongest in the vicinity of Spring Lake, was borne to shore on an east-southeast wind from the ocean. The first reaction of residents of homes equipped with oil furnaces was that something had gone wrong with their heating. However, "when investigation disclosed this was not the case, police headquarters in the area were flooded with calls to learn the cause," a newspaper noted. The cause of the odor was not determined, and the incident was added to the list of nerve-wracking events that exacerbated a general sense of anxiety among civilians.[133]

In another upsetting and bizarre occurrence in early February, a whale with large holes in its body washed up on the shores of Long Beach Township. Authorities said the creature was either struck by a depth charge, torpedoed by an enemy submarine or struck by a mine. Several local residents expressed the opinion that "some American aircraft possibly used the whale as a target while on coastal patrol or might have suspected the creature was a U-boat," according to a local news account.[134]

In addition to the fear of bombing and invasion, another severe challenge of those early war days was the endless procession of shortages and rationing measures. While newspaper headlines screamed the latest dire developments in the Pacific theater —"Singapore Totters Under Jap Assault," read a February 10 headline—New Jersey housewives struggled to prepare nutritious meals with "ersatz" ingredients and planned to grow their own vegetables that summer in tiny backyard "Victory Gardens." At luncheonette counters, sugar bowls had disappeared, and customers were allowed one or two lumps of that suddenly precious commodity for their coffee or tea.

As a child growing up during those austere times, Monmouth County's Judith Stanley recalled in a 2004 oral history interview that she "just didn't have the choice of food that children have today." Consumers became accustomed to dietary substitutes such as oleo for butter, and young Judith "maybe had a Coca Cola once a week or cookies once a week."[135]

Cutbacks in civilian services were constant, from a reduction in the frequency of home dairy deliveries (during an era when the milkman delivered milk, butter and eggs to most homes) to cutting back on public

The 1942 German submarine offensive targeted oil tankers. By the fall, oil rationing was in full force, and this Standard Oil photo caption reads, "Make it a point to go to the thermostat and turn it down ten or fifteen degrees before you leave the house for any length of time, just as this housewife is doing before she goes out for an afternoon of bridge." *Joseph Bilby.*

utility services. On February 8, Jersey Central Power and Light announced that because of the "drastic priorities" placed on war materials, customer services would be limited until "world conditions improve." Meter readers, service representatives and other workers were restricted to a reduced number of daily calls, and whenever possible, employees were expected to travel to their customers' homes on bicycles due to fuel rationing and vehicle shortages.[136]

Ironically, although resources were scarce, many civilians were feeling flush thanks to the higher wages they were earning in defense industries, as full employment finally ended the aftereffects of the Depression. New Jersey shore waitresses reported that "despite the higher prices on menus the public was definitely treating itself to finer foods," often opting for steak over hamburger. One news report attributed the craving for more expensive eats to "a condition of nervous tension brought on by the war."[137]

Certainly, during those dreary early months of 1942, the situation appeared grim, and even the "women's pages" of magazines and newspapers were filled with a spirit of duty and stoicism. *Asbury Park Press* advice columnist Kathleen Norris urged her New Jersey readers to "get busy" and "forget your petty worries and trials." Instead, women needed to devote their lives to "prayer and service," she noted. "We have to do that now," Norris wrote. "If we don't, we may lose the essential that that has made us American and the world a free world. We stand in serious danger; there is no room anymore for trifles."[138]

Perceiving the need to maintain a sense of optimism and purpose, state officials worked to allay the public's fears. An issue that hit especially close to home was local concern about the prospects for Jersey Shore tourism in the coming summer of 1942. One of the key concerns of shore businessmen was that nervous would-be visitors would cancel their vacation plans because of fuel shortages and fears of Axis attacks, perhaps launched from prowling submarines, on the seemingly vulnerable beaches. Much was at stake, since tourism was an enormous industry that drew visitors from throughout the region in peacetime. "Like China, New Jersey absorbs the visitor," a 1939 travel guide noted. "On summer weekends, when city asphalt is soft enough to take heel prints, the State's highways are thronged with the cars of New Yorkers and Pennsylvanians bound for the coast resorts."[139]

As early as March, shore area mayors were bombarded with calls from the owners of summer homes asking if it was true they would not be permitted to open their cottages or go to the beach that summer. Some businessmen, such as Long Branch West End realtor William E. Kirsch, suggested that officials take some immediate action to "combat hysteria, which might adversely affect the Shore."[140]

Many of those with a stake in New Jersey tourism were quick to refute such concerns. Belmar mayor Leon T. Abbott said there was no need to think that the number of summer visitors would be seriously curtailed by "fear and hysteria resulting from the war." In fact, Abbott told a reporter, an unusual number of summer visitors to the shore were seeking permanent homes there. The general feeling among the pubic, he noted, was that "there is a greater safety in the Shore's sparsely populated areas than in the cities, the more logical places of enemy action." How many people he convinced with his pitch is unknown.[141]

Trenton officials announced their plan to hold a series of conferences in February and March to tackle the "warborn recreational problem." Topics to be addressed at the regional meetings included finding ways to attract tourists in spite of tire rationing and gasoline conservation programs, partly by providing more bus and rail transportation.

Arlyn W. Coffin, managing director of the New Jersey Council, struck a patriotic note when he called on New Jersey's "recreational interests to do their part." "As the war proceeds, men and women will be working longer hours, more strenuously and with greater responsibilities," he proclaimed at a press conference. "To New Jersey will be entrusted the revitalization of those tired executives and workers, in order that they may maintain their best efforts and obtain maximum production until victory is achieved."[142]

Despite the uncertain times, tourism officials remained unblinkingly, if perhaps unrealistically, optimistic about the coming summer season. W. Bruce MacNamee, chief of the United States Travel Bureau, predicted that Americans would flock to New Jersey's seaside resorts in greater numbers than ever before. MacNamee urged merchants, home owners and concessionaires to "drop the long faces they have worn since the Pearl Harbor attack." He even predicted blackouts would actually help business by forcing vacationers into "taprooms, restaurants and amusement palaces where they will spend money."[143]

Such sunny predictions, however, were tempered by vacationers' concern that the shore was indeed not worth visiting during a time of fear and uncertainty, when rumors ran wild. As the summer season approached, tourism officials told state leaders that shore merchants deserved "considerable advertising" that would be necessary to "offset the harmful publicity which had previously gone out as to Shore conditions."[144]

One of those unpleasant aspects of publicity was an environmental consequence of the offshore oil tanker sinkings—the accumulation of oil, tar and debris on formerly pristine Jersey beaches from Cape May to Sea Bright. Unfortunately, the morning light would often reveal "a beach covered not with clamshells and seaweed but with burned wood, pools of fuel, and twisted metal—the grim debris of a ship that had been attacked and sunk during the evening." Beachgoers became accustomed to cleaning their soiled feet with kerosene after a day at the beach because of the "tar balls" that were sometimes an unpleasant aspect of a session on the sand.[145]

The cost of cleaning up the mess became a source of contention for beleaguered resort towns; in early June 1942, for example, Governor Edison was informed that the mayors of coastal towns from Belmar to Manasquan were alarmed by the cost of removing oil and tar that had washed up on their shores. A meeting between state and town officials resulted in a joint cleanup effort that buried as much of the residue as possible and removed the rest. The task was time-consuming and labor intensive, involving some three hundred men employed by the WPA. By late June, the job was 50 percent complete, and the beaches were considered clean enough to allow bathing.[146]

Unfortunately, on the July Fourth weekend, state officials reported that once again "considerable oil scum had come in on the beaches" along the Jersey coast and that "some three miles off shore a large pool of oil could be noted." The residual effect of tanker sinkings lasted long beyond the actual date on which they were torpedoed.[147]

Local officials apparently thought the best strategy during such trying times was to plow ahead with business as usual, as evidenced by Sea Girt's beach regulations pamphlet for the 1942 season. The Monmouth County resort's four-page guide made no mention of environmental problems or other wartime issues. Instead, it focused on the typical regulations associated with any beach season. Topics ranged from the cost of seasonal beach badges (three dollars) to the prohibition of "ball throwing or ball games" on the beach during bathing hours.[148]

For young visitors like sixth-grader Jean Thompson Weiler, however, the once welcoming Jersey Shore had acquired a slightly dark and disturbing edge in 1942. Even the enjoyable activity of swimming off the shore of Seaside Park had been altered by the war; Jean recalled that she had to "use a strong solvent [possibly kerosene] to remove tar and oil from her feet before going into her house after swimming."[149]

Unpleasant environmental consequences were just one of the factors that contributed to an uneasy mood. From early 1942, beaches were declared off-limits from sunset to sunrise, and the Coast Guard restricted night fishing and beach parties. Initial patrols by unarmed coast guardsmen were not taken seriously, until the arrest of four Nazi saboteurs landed by a German submarine on Long Island in June 1942. The patrols quickly became more serious and heavily armed, and an army detachment moved in to camp on the Sea Girt Lighthouse lawn, as the New Jersey National Guard's 113th

Infantry Regiment, which had been called to active duty in September 1940 and stationed at Fort Dix, was assigned to guard beaches from Long Island to Delaware Bay.[150]

Lighthouses like Sea Girt's provided ready-made lookouts and were supplemented by observation towers erected along the beaches, all manned twenty-four hours a day. These towers "may have been fine for scanning the sea for enemy submarines, but they were none too pleasant as places to pass the time." In a 1990 newspaper interview, a former coast guardsman recalled his time in the Avalon tower: "We had no power and to keep warm you had to pull coal up from below in a bucket on a string to feed the coal furnace. The most modern thing at Avalon was a big batch of kitchen matches and a candle. Once you lit the candle and made your log entry, you had to put the candle out because of the blackout on during the war."

Beginning in September 1942, mounted coast guardsmen, riding former cavalry horses with saddles issued by the Army Remount Service, along with dog handlers who "patrolled the beaches with German shepherds, on the alert for spies attempting to land from Nazi U-boats," were a regular feature of evenings at the shore, while "observation blimps hovered overhead, ready to drop depth charges on enemy submarines." Jean Thompson Weiler recalled that, through a child's eyes, encountering the beach patrollers on horseback was unsettling: "You'd walk up to the beach and everything would be pitch black…All of a sudden, in the distance, you'd hear hoofbeats…It was really a very creepy thing the first time."[151]

Adding to the sinister undercurrent of wartime was the "dim-out" along the Jersey Shore, authorized in March 1942 and designed to deter enemy attack by eliminating possible silhouetting of passing ships by shore lighting. The visual effect was sobering: "From the ocean the usually bright, cheery boardwalks looked like ghost towns as blackout curtains cut off (or at least reduced) as much light as possible from the concession stands."[152]

Efforts to uniformly enforce dim-out regulations along the Jersey coast became a source of controversy and disagreement, reflected in the copious minutes of Governor Edison's War Cabinet, which was convened weekly during the war by Edison and his successor, Walter Edge. In March, both the navy and the army's Second Corps had requested the dim-out, noting that they "desired all lights along the NJ seacoast be dimmed or eliminated as the preset lights silhouetted a ship at sea making it an easy target for submarines lurking beyond the sea lanes where ships travel."[153]

Unfortunately, throughout the rest of the year, the army and navy were ambivalent, unclear and often contradictory as to which lighting—

New Jersey governor Walter "Wally" Edge, who had been governor during World War I, took office for the second time in January 1944. We see him here with fifteen-month-old Harolyn Cheryl Meyer of Newark, who seems a little apprehensive as he congratulates her on election as "pin up queen" of the USS *New Jersey* in March 1945. *Joseph Bilby.*

boardwalk, inland business, street and home or automotive—they wanted "dimmed or eliminated." State officials often ended up clashing with the military, frustrated by the inconsistent directives. For example, a June 1, 1942 survey of the shore from New York to Long Branch by state authorities, including state police superintendent Colonel Charles H. Schoeffel and Civil Defense director Leonard Dreyfuss, revealed that "the lighting in New Jersey did not in any way reflect a glare on the waters which would silhouette a ship." They added that a visit to Asbury Park found that city completely blacked out. Although the army's Second Corps Command agreed, the navy contended that the dim-out was "not sufficient" at Wildwood Crest and Atlantic City. In return, the state noted that, as observed before, the naval station at Cape May provided the exception to an otherwise satisfactory situation, as its lights could be seen far out at sea. A dredge working at the base was described as being "lit up like a Christmas tree."[154]

Dim-out and blackout requirements, the latter called for in inland areas as a protection against potential air attacks and practiced in drills even before American entry into the war, also clashed with industrial and recreational business needs. As one of the most industrialized states in the United

Colonel Charles H. Schoeffel, New Jersey state police commander. *New Jersey State Police.*

States, New Jersey was a vital cog in the "arsenal of democracy," building military and merchant ships, vehicles, aircraft and ammunition. In April 1942, Harrison's Crucible Steel Company indicated it could not completely blackout operations during a drill or unlikely actual air attack because of "large furnaces where the light could not be extinguished or screened." The following month, army officials expressed their dissatisfaction with the efforts of the owners of Palisades Park, a popular amusement center along the Hudson River on the heights of Cliffside Park and Fort Lee, to comply with the dim-out mandate by operating at 22 percent of normal capacity. Company officials felt that any further dim-out would "put them out of business." Similarly, the operators of night baseball game stadiums indicated they would be forced to shut down completely under the rigorous requirements.[155]

What the dim-outs and blackouts did produce, according to New Jersey officials and Mayor LaGuardia of New York City, was an increase in vehicle accidents and a spike in crime. Automobiles were restricted to using parking lights only (later on, headlights were allowed to be partially masked) while night driving in coastal dim-out areas. A subsequent survey revealed that in Cape May County, where there had been no fatal automobile accidents in 1941, eleven people died in car crashes between January and October 1942. Ironically, in Cape May, most of the tickets issued by local police for having headlights on were issued to navy officers.[156]

Right: The Federal Shipbuilding and Drydock Company in Kearny, New Jersey, continued to turn out both military and merchant ships throughout the war. *Joseph Bilby.*

Below: New Jersey governor Charles Edison (right) listens as Newark mayor Vincent J. Murphy shows him the locations for sirens and whistles to signal a blackout in the state's largest city. *Joseph Bilby.*

After months of wrangling between cabinet and military officials, Governor Edison called for a spirit of cooperation at a July 7 meeting: "The Governor felt that nothing should be done to antagonize the military officials, that their orders should be obeyed, the dim-out regulations enforced and that requests for review on the present restrictions should be brought about in dignified manner and without undue pressure."[157]

The dissension was about to abate anyway, since by July, more effective antisubmarine war tactics began to take effect; by September, the Germans' "Happy Time" of indiscriminate attacks on offshore shipping along the New Jersey coast was effectively over.

Yet even during those early days of '42, when prowling Nazi submarines symbolized the enemy's capability to threaten homeland security, Jersey shore residents managed to find diversions and block out the bad news, at least for a little while. During those dark days, young people like Julia Rifici, a lifelong Asbury Park resident, distracted themselves with ice skating at the Asbury Park Casino and ogling the various celebrities—including Clark Gable and Rudy Vallee—who visited the city. Newspapers devoted extensive coverage to diverting events such as the annual dog show in New York. In February 1942, Monmouth County sent some of its "best canine blood" to Madison Square Garden's annual Westminster Kennel Exhibition, where huge Great Danes mingled with midget Mexican hairless dogs.[158]

In addition, numerous events were held to raise the spirits and stir the patriotic fires of a beleaguered public. "Thrilling! Inspirational!" raved the headlines for a Red Cross patriotic rally held on February 23 at Convention Hall in Asbury Park. The eclectic agenda included a "massing of the colors," a performance by "internationally known dancer" Ruth St. Denis and a broadcast of President Roosevelt's speech to the nation. Asbury Park merchant and Bradley Beach resident Harry Whelan established a "Dad's Club" for the families of soldiers and sailors on the Bradley Beach boardwalk.[159]

That patriotic spirit was also inspired by the increasing presence of young soldiers preparing for or recovering from battle, who were especially prevalent in Atlantic City, known as "Camp Boardwalk." The city, with its wide beaches and broad, even boardwalk, was considered an ideal location for military training, and beginning in 1942, the resort was flooded with thousands of boisterous GIs of the army air force readying for war, who stayed in forty-seven hotels and hostels occupied by the military.

The change in the urban landscape was dramatic. For years, Atlantic City had been known as an "amusement factory," a "carnival city as characteristic of this country's culture as a Brighton or the Riviera are of Europe's." The boardwalk's "glittering, luxurious front" was lined by shops, restaurants, exhibit rooms, booths, auction houses, an occasional bank and even a private park. The beach was known for its "miles of white sand," "dotted with cabanas, umbrellas, beach chairs, and gaily togged bathers." Abruptly, the city shifted to a wartime stance: "Instead of tourists, some half-million servicemen were billeted in the hotels and marched and drilled on the Boardwalk."[160]

On the beach, soldiers participated in training exercises that prepared them for fighting in Europe and the Pacific. The boardwalk provided an even path for wounded soldiers going through physical therapy, particularly those adjusting to prosthetic limbs. Boardwalk landmarks were renamed to reflect the martial tenor of the times; Convention Hall in Atlantic City, for example, became the Army Air Corps Technical Training Command Center.

Despite the austerity of wartime conditions, with lights out at night and windows covered with blue cellophane, the city was infused with a youthful energy that countered the grim war news bombarding the American public. The wartime excitement was heightened by the presence of numerous celebrities who came to entertain military personnel, including Bob Hope, the Andrews Sisters and Joe DiMaggio. In an effort to maintain some sort of propriety during such heady times, city fathers prohibited young girls under eighteen from walking the boardwalk unescorted after 9:00 p.m.

The freewheeling atmosphere also created new social opportunities for gay servicemen and women, who, freed from the constraints of narrow-minded hometowns, were eager to explore a booming bar scene. Men in search of other men wandered to the Entertainer's Club and several other New York Avenue bars featuring drag shows. Other bars began to welcome gay clientele, and guesthouses catering to homosexuals opened up nearby: "The development of these homosocial spaces was done by word of mouth" as opposed to advertising "for fear of police or extralegal retribution." The city's long established brothels and illegal casinos, needless to say, prospered as well.[161]

For the young and inexperienced, however, Atlantic City during wartime was the place for innocent fun. Teenager Alyce Crowe, who took a waitressing job at an Atlantic City hotel in 1942, recalled her time at the resort with fondness: "In the evening after work, there were lots of dates and walks on the Boardwalk, dances on the Steel Pier, amusements on the Million Dollar Pier, and the wonderful Old Time Movie House on the Walk which showed silent films."[162]

Asbury Park was also transformed by the invasion of military personnel. Before the war, the city had been regarded as "one of the best-known resorts in northern New Jersey," with a boardwalk lined with "eating places, a fishing pier, recreational attractions, solariums, and shops where everything from imported Oriental rugs to souvenirs of the *Morro Castle* disaster are sold." As with Atlantic City, the city's festive ambiance took on a distinctly martial tone once war was declared. Two of the city's signature hotels—the Berkeley Carteret and the Monterey—were set aside for rest and relaxation for Allied forces, including members of the British navy. Like its neighbor to the south, Asbury Park became a hub of military activity: "Sixth Avenue between the two hotels served as the parade ground for morning drills," and most nights, "the Salvation Army canteen came to the main gate at the Berkeley to give out cocoa."[163]

Even sleepy Sea Girt was infused with a dose of wartime energy when the army signal corps leased property of the National Guard camp located in the quiet shore town. The Sea Girt base—with mess hall space for 1,700 men, a post exchange and other buildings as well as its outstanding rifle range—provided a ready-made training site. Recruits drilled on the parade ground, fired on the ranges and conducted route marches up and down nearby highways and out

New Jersey National Guardsmen on maneuvers on the Asbury Park beach in 1942. *Joseph Bilby.*

into what is now Allaire State Park. The routine of daily military exercises was enlivened by on-base USO dances, where a GI band belted out hit tunes while soldiers jitterbugged frenetically with local girls.

Shore residents also witnessed actual military maneuvers. The Forty-fourth Division, a largely New Jersey Army National Guard unit, had been called to active duty at Fort Dix in November 1940 and participated in a large training exercise in May 1942, in which forces were deployed all along the New Jersey coast to repel a mock invasion.

Inland New Jersey also did its part. The United Seaman's Service and the War Shipping Administration teamed up to create a "rest center" at Gladstone for "convalescing seamen" whose ships had been torpedoed off the coast or on convoys to Europe. The center was located on the five-hundred-acre estate of Charles and Mary Suydam Cutting, wealthy international celebrities. Charles was an explorer and the first westerner to

The United Seaman's Service and the War Shipping Administration teamed up to create a "rest center" at Gladstone, New Jersey, for "convalescing seamen" whose ships had been torpedoed. On May 31, 1943, the seamen in residence were visited by the Duke and Duchess of Windsor, who chatted with them and signed autographs, as seen in this photo. The caption described the duchess's "colorful" dress as "inscribed with gin rummy terms." *Joseph Bilby.*

enter the "forbidden city" of Lhasa, Tibet. On May 31, 1943, the seamen in residence were visited by the Duke and Duchess of Windsor, who chatted with them and signed autographs. The newspaper story was accompanied by a photo caption that described the duchess's "colorful" dress as "inscribed with gin rummy terms."[164]

Youth, energy, optimism—these were some of the factors that kept homefront morale high along the New Jersey coast during the early days of the war, when victory was far from certain. In spite of the ominous presence of stealthy U-boats that took numerous lives and a turbulent world situation that endangered national security, citizens at the Jersey shore carried on, dutifully complying with blackouts, dim-outs, rationing and the other onerous civic duties of wartime.

As Julia Rifici said of growing up in Asbury Park during that trying era: "We had a lot of different things we had to do to 'help the cause,' as they said." And they did what they had to do—and did it well.[165]

Opposite, top: By the time this snapshot was taken looking seaward at Ocean Grove, New Jersey, in the summer of 1944, the submarine threat was over. *Joseph Bilby.*

Opposite, bottom: At the end of the war, the Federal Shipbuilding and Drydock Company in Kearny went to work removing deck guns from merchant ships. *Joseph Bilby.*

Chapter 5
AFTERWARD

"You Had to Bang on the Bar to Get His Attention"

In 1975, when he was just out of college, Jay Amberg was freelancing as a reporter for several weekly papers along the Ocean County, New Jersey shore, writing articles for outdoor publications and doing a lot of fishing on Long Beach Island. As a local journalist, Amberg would occasionally stop at the Barnegat Light Coast Guard Station to see if there was a newsworthy story there, and there often was. One summer day, Jay was at the station when a call came in from the *Snoopy Gus*, a fishing trawler owned by Gus Brindley. Brindley, who was trawling for butterfish about twenty-five miles directly off Barnegat Light, told the coast guardsmen that he thought he had snagged a torpedo in his net.[166]

The station commander dispatched a forty-four-foot Motor Life Boat to the rescue, and Jay hopped aboard, as there was definitely a story in Brindley's catch. About an hour and a half later, the Coast Guard boat pulled up alongside *Snoopy Gus*, and Jay immediately spotted what was clearly, although somewhat rusted in its center, a torpedo suspended in a net off the boat's aft. The coast guardsmen radioed the situation back to Barnegat Light and were ordered to stand by as an Explosive Ordnance Disposal (EOD) team was dispatched to the scene from Naval Weapons Station Earle in Monmouth County.[167]

The sailors arrived by helicopter shortly afterward, rappelled down to the trawler and ordered Brindley and his mate, "Dragger Joe" to board the Coast Guard boat as they dealt with the torpedo. The EOD men cut the net and removed it, along with the torpedo it held, from the boat and attached a buoy to keep it afloat. They told Brindley and Joe to get back on their fishing boat and ordered both boats to leave the area. When they were about a half mile away, they heard an explosion. Jay turned and saw a massive plume of water erupt skyward. The results of the war on the Jersey shore had echoed for more than thirty years after it ended—and they still do.[168]

They echo most loudly within the shore's commercial fishing community. Captain Eric Olsen, a longtime fisherman and mayor of Barnegat Light, recalls a time back in the early 1980s when he was dragging for scallops out near the *R.P. Resor* wreck and came up with a five-hundred-pound bomb in his net, which he quickly cut loose, accepting the loss of expensive gear rather than taking a chance with his and his crew's lives. On another occasion, Olsen was scallop fishing off Long Island and trawled up a torpedo and again took the net loss rather than the risk.[169]

Olsen and his fellow captains were very familiar with the tragic tale of the *Snoopy*, a commercial fishing boat that trawled up a bomb or torpedo off the North Carolina coast back in July 1965 while fishing in "Torpedo Alley." The crew brought the rusty ordnance on board *Snoopy* and tried to remove it from their mesh scallop bag, but it exploded and killed eight of the boat's twelve crewmen. In the immediate aftermath, the Coast Guard advised captains to simply cut their nets and losses when they came up with a bomb or torpedo.[170]

Along with unexploded ordnance, the war along the New Jersey coastline left an enduring treasury of tales in its wake, including the story of the Barnegat bartender, a World War II merchant mariner who, after his ship was torpedoed, drifted alone in a lifeboat in freezing weather before he was rescued. His ears clogged with ice, which cost him his hearing. "He would keep his hands on the bar and you had to bang on it to get his attention. He could feel the vibration and then would turn and read your lips for the drink order," Eric Olsen recalled. Another tale from the 1960s involves a commercial fishing boat crew that trawled up a torpedo and brought it into shore near the Manasquan Coast Guard Station and started disassembling it to get metal to sell for scrap, until coast guardsmen happened to notice and evacuated the area.[171]

Other than the tangible recollections of commercial fishermen, few people alive today remember the submarine war off the New Jersey coast. Local legends based in hazy reminiscences have long persisted, however, and folklore related to the era abounds along the Jersey shore. These stories are reminiscent of Little Joe's binocular sighting of what he thinks, drawing on

Biff Baxter's almost visual verbal radio description, is a German submarine off Rockaway Beach in Woody Allen's wonderful film *Radio Days*. Author Joseph Bilby has heard similar tales from local old-timers, and while entertaining and charming, they should be taken with more than the usual proverbial grain of salt.[172]

When John Chatterton was trying to gather evidence to discover the identity of the "Mystery U-boat" he had uncovered while diving off Point Pleasant in 1991, he drafted a press release that led to an article in the *Newark Star Ledger*, the state's largest circulation newspaper, that resulted in a number of rather dubious contacts, including "sons, mothers, brothers and grandchildren [who] swore that loved ones had attacked and sunk a U-Boat on a secret mission that the government still refused to acknowledge." The assertion that Germans had actually landed on the New Jersey shore from U-boats is a legend that has never really died, and Chatterton heard from "others [who] told of seeing U-Boat crewmen swimming onto American shores to buy bread or attend dances." He even got an offer to "buy a Nazi skull" should he come up with one in a dive.[173]

But some shore residents also have real, if not so dramatic, memories of the era, and although they are fading with the years, some endure. Allen Atheras, a young Spring Lake resident in the summer of 1942, remembered the town's beaches covered with oil and debris that year. His mother left a jar of kerosene at the back door for him and his sister to wipe the beach sludge off their feet. On one occasion, Atheras rummaged through a battered and beached lifeboat, found a Model 1911 .45-caliber automatic pistol and brought it home as a souvenir. A .30-caliber Lewis machine gun in the collection of the National Guard Militia Museum of New Jersey in Sea Girt allegedly came ashore at Manasquan in a lifeboat during the summer of 1942, but there is no paper provenance for the story.[174]

Lieutenant Margaret Jennings of Spring Lake Heights had a memorable encounter with a German submarine far from home. Jennings, who was a graduate of Saint Rose High School in Belmar and Saint Vincent's Hospital nursing school in New York, was an army nurse assigned to the *Seminole*, a hospital ship returning from the Mediterranean war zone. She recalled that "a German U-boat watched her ship at periscope depth during a burial at sea" before moving on to attack a convoy. Luckily for Jennings and the crew and patients on the *Seminole*, the U-boat commander obeyed the rules of war and refrained from launching a torpedo at the hospital ship.[175]

The submarine war of 1942 has also survived in local folk art. One example is a hand-carved copy of a German submarine crafted a half

Jersey shore submarine folk art. This hand-carved copy of a German submarine was made a half century ago by William C. Wienkop for his daughter Marjy. During the war, Wienkop lived in Union, New Jersey, but his family had a summer home in Point Pleasant. *Marjy Weinkop.*

century ago by William C. Wienkop for his daughter Marjy. During World War II, Wienkop lived in Union, New Jersey, but his family summered on the Metedeconk River. He served as a lieutenant junior grade Coast Guard communications officer in the early 1950s, working on vessels patrolling off New England and up to Greenland, no doubt alongside coast guardsmen who were veterans of the Jersey shore war. Wienkop moved to Point Pleasant after his marriage and shared submarine stories with his children. Eight-year-old Marjy painted the numbers on the sub after asking her father for the designation of a famous U-boat. *U-505* was a U-boat captured by the U.S. Navy and is one of the few surviving German World War II submarines. It is on display at the Chicago Museum of Science and Industry.[176]

U-boats did come very close to the American shore on occasion, although not in New Jersey, and the stories of their missions reverberated in the Garden State. On the night of June 13, 1942, four German saboteurs and a supply of explosives and priming devices were landed on a beach near Amagansett, Long Island, from *Kapitanlieutenant* Hans Heinz Linder's *U-202*,

Above: A coast guardsman on patrol duty along the New Jersey coast. After German saboteurs who landed on Long Island and in Florida were captured, coastal defense intensified. *Joseph Bilby.*

Right: Anthony Cramer was arraigned in a federal court in Newark, New Jersey on July 17, 1942, for the crime of treason. Cramer, a World War I veteran of the German army, immigrated to the United States in 1925 and later became a naturalized citizen but assisted the German saboteurs landed by submarine in 1942. *Joseph Bilby.*

which went on to sink the *Rio Tercero* off the New Jersey coast. Four days later, another submarine landed four more Germans on Ponte Vedra Beach, near Jacksonville, Florida, on a similar sabotage mission.

All of the saboteurs, part of what was called Operation *Pastorius*—named, ironically, after Franz Daniel Pastorius, the founder of Germantown, Pennsylvania, the first permanent German settlement in America—were captured, as well as sixty-five Americans, mostly German immigrants, who aided them in one form or another, and there were several New Jersey connections, including New York resident Anthony Cramer, who was arrested in Newark. Cramer, a World War I veteran of the German army, immigrated to the United States in 1925 and later became a naturalized citizen. He was a personal friend of one of the saboteurs, Werner Thiel, and met with Thiel and an accomplice, Edward Kerling. Thiel gave Cramer a money belt with several thousand dollars in it for safekeeping.[177]

Cramer was arraigned in a federal court in Newark on July 17, 1942, for the crime of treason, as he was a U.S. citizen. He was subsequently convicted and, although escaping the death penalty the prosecutor asked for, was sentenced to forty-five years in prison.[178]

Cramer appealed his sentence, and in a 1945 5–4 decision, the Supreme Court reversed the conviction, maintaining that the criteria for treason—two witnesses to an overt act of treason, not just hanging out with potential saboteurs and watching one's bankroll—was not met. He subsequently entered a guilty plea to the charge of trading with the enemy and violating a "federal freezing order of enemy funds" and was sentenced to six years in prison.

The final arrest in the *Pastorius* plot was that of Carl Emil Ludwig Krepper in Newark. Krepper was born in 1884 and immigrated to America from his native Germany in 1909 as a theology student. Krepper had studied in the Kropp Lutheran Seminary in his homeland, an institution designed to provide pastors for the German American church. He transferred to the Lutheran Theological Seminary at Philadelphia for a year of final preparation before being ordained and assigned to a Lutheran church in Pennsylvania or New Jersey. Krepper filed citizenship papers in 1919 and was granted citizenship in 1922. During World War I, a number of his fellow Kropp students in America were accused of attempting to undermine the war effort, but not Krepper, who registered for the draft, although he was not conscripted.[179]

In 1923, Krepper was transferred from his parish in Philadelphia to become pastor of two Lutheran churches in Rahway and Carteret, New Jersey, and would spend the rest of his clerical career in the state. In 1932, he was appointed pastor of the First German St. John's Evangelical Lutheran

Church in Newark. Krepper was active in church activities but began, around this time, to take trips back to his native Germany. He became active in the German-American Business League (DAWA), an organization that campaigned against proposed boycotts of German goods and for boycotting Jewish-owned businesses, as well as the German-American Bund organization of Nazi sympathizers, which maintained several bases in New Jersey, most notably Camp Nordland in Sussex County. In 1935, he took a two-year leave of absence and then resigned from his parish to make several long trips back and forth to Germany, where he met Walter Kappe, a German intelligence officer who recruited him to "do propaganda work for

Carl Emil Ludwig Krepper of Newark was the principal contact man for the German saboteurs landed at Amagansett, Long Island, by *U-202* in 1942. *Joseph Bilby.*

Germany in the United States and provide funds and haven for Bundists persecuted by Jews" in New Jersey.[180]

He did far more than that. Leaving his wife behind in Germany, Krepper returned to America for a final time via Portugal on December 16, 1941, and was soon deeply involved in Operation *Pastorius*, assigned as the American contact to provide safe houses and money for the German saboteurs landed from submarines on Long Island and in Florida. No longer a pastor, Krepper was working as a bookkeeper for the Downtown Club in Newark and living in a rooming house at 68 James Street in the city. The saboteurs tried to contact Krepper and failed and, before they could make another attempt, were captured.[181]

The FBI caught on to Krepper, put him under surveillance and caught him in a sting operation in 1944. He was charged with a variety of espionage-related crimes and housed in the Hudson County Jail in Jersey City until his trial in February 1945. Convicted, Krepper was sentenced to twelve years in jail. Attempts to appeal his conviction were denied in 1947 and 1950, and after his release he faded into history. Carl Emil Ludwig Krepper died at a nursing home in Massachusetts in 1972, and his body was cremated. No one claimed his ashes.[182]

The New Jersey shore U-boat story that gained the most publicity in the postwar era was that of the long search to identify the German submarine found by scuba divers John Chatterton and Richie Koehler off the Jersey coast in 1991. German and American naval sources gave no clue as to the submarine's identity, but after seven years of research, much of which led to dead ends, Chatterton and Koehler succeeded in verifying that it was *U-869*, which was supposed to be operating off the coast of Africa when it disappeared in early 1945.

Although it did not date from the "Happy Time" of 1942 and did not sink any ships off the New Jersey coast, the fact that it met its end off Manasquan made *U-869* a New Jersey story. Robert Kurson's best-selling book *Shadow Divers* and a subsequent public television *Nova* show publicized the discovery and eventual identification of *U-869*. Chatterton and Koehler concluded that the submarine in effect sunk itself with a malfunctioning torpedo that turned around and hit it. That account has been contested by a Coast Guard assessment and in Gary Gentile's 2006 book *Shadow Divers Exposed: The Real Saga of the U-869*. Gentile, a noted wreck diver and author himself, cites logs and crew accounts to posit that *U-869* was, as the Coast Guard concluded, damaged and then destroyed by depth charges fired by the American destroyers USS *Howard D. Crow* and USS *Koiner*. And so the stories of the submarine war off the New Jersey shore have endured and evolved and will, no doubt, continue to do so.

NOTES

CHAPTER 1

1. Lake, *Submarine in War*, 77–79.
2. Ibid., 79–80.
3. Ibid., 81–83.
4. Veit, "Innovative, Mysterious *Alligator*."
5. Ibid.
6. Ragan, "Union Whale Surfaces."
7. Ibid.
8. Bilby, ed., *New Jersey Goes to War*, 57.
9. *New York Times*, September 19, 1884; Lake, *Submarine in War*, 83–84.
10. Morris, *John P. Holland*, 13, 15.
11. Lake, *Submarine in War*, 89.
12. Whitman, "John Holland"; Lake, *Submarine in War*, 95–96.
13. Morris, "John P. Holland"; Whitman, "John Holland"; Lake, *Submarine in War*, 93–94.
14. Morris, "John P. Holland"; Whitman, "John Holland"; Lake, *Submarine in War*, 94.
15. *Boston Journal*, August 7, 1883; *New York Herald*, April 13, 1883; Whitman, "John Holland."
16. *Maysville (KY) Evening Bulletin*, May 31, 1883; Whitman, "John Holland."
17. Whitman, "John Holland."
18. *Brooklyn Daily Eagle*, April 19, 1907.
19. Morris, *John Holland*, 47.
20. Morris, "John P. Holland"; *Brooklyn Daily Eagle*, August 13, 1914.

CHAPTER 2

21. State of New Jersey, *Record of Officers*, 140; Salter, *History of Monmouth*, 291; Salter and Beekman, *Old Times*, 149.

22. Dorwart, *Cape May County*, 75–76.

23. Lee, *New Jersey as a Colony*, 99; Strum, "South Jersey," 9–13.

24. Barber and Howe, *Historical Collections*, 70, 363; Powell, *History of Camden County*, 77–78; Lee, *New Jersey as a Colony*, 101.

25. Federal Writers Project, *New Jersey*, 73–74; Lender, *One State in Arms*, 76, 78.

26. Bilby, Madden and Ziegler, *Hidden History of New Jersey at War*, 69–74; Federal Writers Project, *New Jersey*, 275.

27. Ibid., 75–76.

28. Ibid., 76.

29. Ibid., 76–78.

30. Gray, *U-Boat War*, 232; United States Navy, *German Submarine Activities*, 17–18, Williamson, *Boats of the Kaiser's Navy*, 15–16.

31. Williamson, *U-Boats of the Kaiser's Navy*, 15–16.

32. Ibid., 11; United States Navy, *German Submarine Activities*, 21–22.

33. United States Navy, *German Submarine Activities*, 7; *New York Times*, February 17, 1917.

34. Ibid.; Roberts and Youmans, *Down the Jersey Shore*, 238.

35. *New York Times*, February 7, 8, 1918.

36. Ibid., January 15, 1918; Federal Writers Project, *New Jersey*, 206.

37. Ibid., March 12, 1918.

38. Ibid., July 12, 1918.

39. Gray, *U-Boat War*, 232.

40. Ibid.; United States Navy, *German Submarine Activities*, 23.

41. United States Navy, *German Submarine Activities*, 24.

42. Buchholz, *New Jersey Shipwrecks*, 140, 142.

43. Thomas, *Raiders of the Deep*, 297.

44. Ibid., 304–05.

45. United States Navy, *German Submarine Activities*, 26–27, 127.

46. Ibid., 31.

47. Ibid., 32.

48. Thomas, *Raiders of the Deep*, 309–10; Buchholz, *New Jersey Shipwrecks*, 142–43.

49. United States Navy, *German Submarine Activities*, 33; Buchholz, *New Jersey Shipwrecks*, 146.

50. *Cincinnati Enquirer*, June 4, 1918; United States Navy, *German Submarine Activities*, 34.

51. United States Navy, *German Submarine Activities*, 35.

52. *Asbury Park Press*, June 4, 1918; Buchholz, *New Jersey Shipwrecks*, 145.

53. United States Navy, *German Submarine Activities*, 35.

54. Ibid., 37; Thomas, *Raiders of the Deep*, 313.

55. *Scranton (PA) Republican*, June 6, 1918; United States Navy, *German Submarine Activities*, 37.

56. Buchholz, *New Jersey Shipwrecks*, 146.

57. *New York Times*, June 5, 1918.

58. Ibid.

59. *Asbury Park Press*, June 3, 1918.

60. Ibid.

61. United States Navy, *German Submarine Activities*, 50.

62. Ibid., 92–93, 132, 134.

63. *New York Times*, June 26, 1918.

64. Report from "CO, Co. F, 22nd Infantry, Subject: Submarine Attack, August 2, 1918," RG 156 entry 5958 box 1 file 052, NA.

65. *New York Times*, October 2, 1918.

66. *Asbury Park Press*, June 3, 1918.

67. New Jersey Scuba Diving Wreck Sites, http://njscuba.net/sites/site_black_sunday.php.

68. Federal Writers Project, *New Jersey*, 51.

69. Thomas, *Raiders of the Deep*, 332–33.

CHAPTER 3

70. *New York Times*, February 28, 1942; Hickam, *Torpedo Junction*, 51.

71. Taylor, *Fire on the Beaches*, 95.

72. *New York Times*, February 28, 1942.

73. Gannon, *Operation Drumbeat*, 174–75; Offley, *Burning Shore*, 20; Hickam, *Torpedo Junction*, 54.

74. Hickam, *Torpedo Junction*, 54.

75. Offley, *Burning Shore*, 37–38; Taylor, *Fire on the Beaches*, 46.

76. Ibid.

77. Taylor, *Fire on the Beaches*, 8–49.

78. Gannon, *Operation Drumbeat*, 214–25, Offley, *Burning Shore*, 53, 107–09.

79. Gannon, *Operation Drumbeat*, 239–41; Offley, *Burning Shore*, 53, 107–09.

80. Gannon, *Operation Drumbeat*, 233–37; Offley, *Burning Shore*, 113–15.

81. *Salt Lake (UT) Tribune*, January 26, 1942; *Altoona (PA) Tribune*, January 26, 1942.

82. *New York Times*, January 26, 1942.

83. Ibid., January 26, 1942; World War II Ships Register, http://www.warsailors.com/singleships/varanger.html.

84. *New York Times*, February 7, 1942; Eastern Sea Frontier Enemy Action Diary, Chapter 5, February 1942.

85. *New York Times*, February 7, 1942.

86. Ibid.

87. Eastern Sea Frontier Enemy Action Diary, February 27, 1942.

88. Ibid.; Morrison, *Battle of the Atlantic*, 134.

89. Eastern Sea Frontier Enemy Action Diary, February 27, 1942; Taylor, *Fire on the Beaches*, 98.

90. Eastern Sea Frontier Enemy Action Diary, February 27, 1942; *Wilkes-Barre (PA) Record*, March 4, 1942; Hickam, *Torpedo Junction*, 60.

91. Eastern Sea Frontier Enemy Action Diary, February 27, 1942; *Wilkes-Barre (PA) Record*, March 4, 1942; Morrison, *Battle of the Atlantic*, 135; Hickam, *Torpedo Junction*, 62–63.

92. Gentile, *Shipwrecks of New Jersey*, 27; Hickham, *Torpedo Junction*, 65; NOAA Screening Level Risk Assessment, 2013. http://sanctuaries.noaa.gov/protect/ppw/pdfs/cayru.pdf.

93. Gentile, *Shipwrecks of New Jersey*, 28.

94. *New York Times*, March 12, 1942.

95. Ibid., March 13, 1942; *Decatur (IL) Daily Review*, March 12, 1942; NOAA Screening Level Risk Assessment, 2013. http://sanctuaries.noaa.gov/protect/ppw/pdfs/cayru.pdf.

96. *Havre (MT) Daily News*, March 13, 1942.

97. *New York Times*, March 12, 1942.

98. Ibid., March 12, 1942; Naval History and Heritage Command, http://www.history.navy.mil/research/histories/ship-histories/danfs/l/larch.html; Eastern Sea Frontier Enemy Action Diary, March 9, 1942.

99. Gentile, *Shipwrecks of New Jersey*, 143–44; *New York Times*, March 18, 1942; Naval History and Heritage Command, http://www.history.navy.mil/research/histories/ship-histories/danfs/l/larch.html.

100. Naval History and Heritage Command, http://www.history.navy.mil/research/histories/ship-histories/danfs/l/larch.html.

101. *New York Times*, March 18, 1942.

102. Ibid.

103. Gentile, *Shipwrecks of New Jersey*, 72–74; Offley, *Burning Shore*, 118–19; Eastern Sea Frontier Enemy Action Diary, March 14, 1942. Gannon, *Operation Drumbeat*, 334–35.

104. Gannon, *Operation Drumbeat*, 344–45.

105. Ibid., 378.

106. Berg, *Wreck Valley*, 10–11.

107. Ibid., 16.

108. Gannon, *Operation Drumbeat*, 346–47; Noble, *Beach Patrol*, 4.

109. Berg, *Wreck Valley*, 100–01; Gentile, *Shipwrecks of New Jersey*, 97.

110. Buchholz, *New Jersey Shipwrecks*, 172–73.

111. Berg, *Wreck Valley*, 100–01; Gentile, *Shipwrecks of New Jersey*, 97.

112. Hickam, *Torpedo Junction*, 229.

113. Gentile, *Shipwrecks of New Jersey*, 114; *New York Times*, June 23, 1942.

114. Gentile, *Shipwrecks of New Jersey*, 114; Eastern Sea Frontier and U-Boat Records, U-202—6th War Patrol, http://www.uboatarchive.net/KTB202-6.htm.

115. Gentile, *Shipwrecks of New Jersey*, 114; Eastern Sea Frontier and U-Boat Records, 6th War Patrol, http://www.uboatarchive.net/KTB202-6.htm; Morrison, *Battle of the Atlantic*, 132.

116. *New York Times*, July 3, 1942.

117. Ibid., July 1, 7, 1942.

118. Ibid., June 26, July 7, 1942.

119. Gannon, *Operation Drumbeat*, 288; Taylor, *Fire on the Beaches*, 216, 223; Morrison, *Battle of the Atlantic*, 257.

120. Hickam, *Torpedo Junction*, 259.

Chapter 4

121. *Asbury Park Press*, January 30, 1942.

122. Ibid., June 3, 1918.

123. Ibid.

124. Lender, *One State in Arms*, 88.

125. Bilby, *Sea Girt*, 101.

126. *Asbury Park Press*, February 15, 1942.

127. Ibid., February 18, 1942.

128. Lurie and Veit, *New Jersey*, 252–53.

129. Buchholz, *Shore Chronicles*, 313–14.

130. Woolf, *4th of July*, 138.

131. Mappen, *Jerseyana*, 210–11.

132. Ibid., 211–12; Lender, *One State in Arms*, 87.

133. *Asbury Park Press*, February 16, 1942.

134. Ibid., February 4, 1942.

135. Aiken, *Remembering the Twentieth Century*.

136. *Asbury Park Press*, February 8, 1942.

137. Ibid., February 15, 1942.

138. Ibid.

139. Federal Writers Project, *New Jersey*, 5.

140. *Asbury Park Press*, January 28, 1942.

141. Ibid.

142. *Asbury Park Press*, February 6, 1942.

143. Ibid., February 15, 1942.

144. New Jersey Governor's War Cabinet Notes, May 26, 1942.

145. Roberts and Youmans, *Down the Jersey Shore*, 100.
146. New Jersey Governor's War Cabinet Notes, June 2, 1942.
147. Ibid., July 7, 1942.
148. Borough of Sea Girt, *Beach Regulations*.
149. Roberts and Youmans, *Down the Jersey Shore*, 100.
150. Bilby, *Sea Girt*, 100–01.
151. Noble, *Beach Patrol*, 15; Buchholz, *Shore Chronicles*, 317; Roberts and Youmans, *Down the Jersey Shore*, 101.
152. Roberts and Youmans, *Down the Jersey Shore*, 192–93.
153. New Jersey Governor's War Cabinet Notes, March 3, 1942.
154. Ibid., June 2, 1942.
155. Ibid., April 28, May 4, 1942.
156. Ibid.
157. Ibid., July 7, 1942.
158. Aiken, *Remembering the Twentieth Century*.
159. *Asbury Park Press*, February 15, 1942.
160. Federal Writers Project, *New Jersey*, 190–93.
161. Simon, *Boardwalk of Dreams*, 161.
162. Buchholz, *Shore Chronicles*, 320.
163. Pike, *Asbury Park's Glory Days*, 126; Federal Writers Project, *New Jersey*, 682.
164. *New York Times*, June 1, 1943.
165. Aiken, *Remembering the Twentieth Century*.

Chapter 5

166. Interview with Jay Amberg, October 27, 2015.
167. Ibid.
168. Ibid.
169. Interview with Eric Olsen, November 5, 2015.
170. Ibid.; *Nashua (NH) Telegraph*, July 26, 1965.
171. Interview with Eric Olsen, November 5, 2015.
172. Bailey, *Reluctant Film Art*, 64–65.
173. Kurson, *Shadow Divers*, 146–47.
174. Interview with Al Atheras, June 25, 2001.
175. Undated newspaper clipping from the Asbury Park Press and oral history interview on file at the National Guard Militia Museum of New Jersey in Sea Girt.
176. E-mail correspondence with Marjy Wienkop, October 26, 2015.
177. *New York Times*, July 14, 1942.
178. Ibid., November 19, 1942; December 3, 1942.
179. Watson and Watson, "Carl Krepper."

180. Ibid.
181. Ibid.
182. Ibid.

BIBLIOGRAPHY

Books

Aiken, Joanna, et. al. *Remembering the Twentieth Century: An Oral History of Monmouth County*. Freehold, NJ: Monmouth County Library, 2002.

Bailey, Peter J. *The Reluctant Film Art of Woody Allen*. Lexington: University Press of Kentucky, 2000.

Barber, John Warner, and Henry Howe. *Historical Collections of the State of New Jersey*. New Haven, CT: John Barber, 1868.

Berg, Daniel. *Wreck Valley: A Record of Shipwrecks off Long Island's South Shore and New Jersey*. Vol. 2. East Rockaway, NY: Aqua Explorers Inc., 1990.

Bilby, Joseph G. *Sea Girt: A Brief History*. Charleston, SC: The History Press, 2008.

———, ed. *New Jersey Goes to War: Biographies of 150 New Jerseyans Caught up in the Struggle of the Civil War, including Soldiers, Civilians, Men, Women, Heroes, Scoundrels—and a Heroic Horse*. Wood Ridge: New Jersey Civil War Heritage Association Sesquicentennial Committee, 2010.

Bilby, Joseph G., and Harry Ziegler. *Asbury Park: A Brief History*. Charleston, SC: The History Press, 2009.

Bilby, Joseph G., James M. Madden and Harry Ziegler. *Hidden History of New Jersey at War*. Charleston, SC: The History Press, 2014.

Birkner, Michael J., Donald Linkey and Peter Mickulas, eds. *The Governors of New Jersey: Biographical Essays*. New Brunswick, NJ: Rutgers University Press, 2014.

Borough of Sea Girt, NJ. *Beach Regulations for the Borough of Sea Girt, New Jersey, Season 1942*. Sea Girt, NJ, 1942.

Buchholz, Margaret Thomas. *New Jersey Shipwrecks: 350 Years in the Graveyard of the Atlantic*. Harvey Cedars, NJ: Down the Shore Publishing, 2004.

————. *Shore Chronicles: Diaries and Travelers' Tales from the Jersey Shore, 1764–1955*. Harvey Cedars, NJ: Down the Shore Publishing, 1999.

Dorwart, Jeffrey M. *Cape May County, New Jersey: The Making of an American Resort Community*. New Brunswick, NJ: Rutgers University Press, 1992.

Edge, Walter E. *A Jerseyman's Journal: The Autobiography of a Businessman, Governor, Diplomat*. Princeton, NJ: Princeton University Press, 1948.

Federal Writers Project. *New Jersey: A Guide to its Present and Past*. New York: Viking Press, 1939.

Fleming, Thomas. *New Jersey: A Bicentennial History*. New York: W.W. Norton and Company, 1977.

Gabrielan, Randall. *Explosion at Morgan: The World War I Middlesex Munitions Disaster*. Charleston, SC: The History Press, 2012.

Gannon, Michael. *Operation Drumbeat: The Dramatic True Story of Germany's First U-Boat Attacks Along the American Coast in World War II*. New York: Harper Perennial, 1991.

Gentile, Gary. *Shadow Divers Exposed: The Real Saga of the U-869*. Philadelphia: Bellerophon Bookworks, 2006.

————. *Shipwrecks of New Jersey*. Norwalk, CT: Sea Sports Publications, 1988.

Gray, Edwyn. *The U-Boat War, 1914–1918*. London: Leo Cooper, 1994.

Hickam, Homer H., Jr. *Torpedo Junction: U-Boat War off America's East Coast, 1942*. Annapolis, MD: U.S. Naval Institute, 1989.

Hinkle, David R., Harry H. Caldwell and Arne C. Johnson, eds. *United States Submarines*. Milford, CT: Universe, 2002.

Johnson, Nelson. *Boardwalk Empire: The Birth, High Times and Corruption of Atlantic City*. Medford, NJ: Plexus Publishing, 2002.

Kurson, Robert. *Shadow Divers: The True Adventures of Two Americans who Risked Everything to Solve One of the Last Mysteries of World War II*. New York: Random House, 2005.

Lake, Simon. *The Submarine in War and Peace: Its Developments and Its Possibilities*. Philadelphia: J.B. Lippincott Company, 1918.

Lee, Francis Bazley. *New Jersey as a Colony and as a State, One of the Original Thirteen*. Vol. 3. New York: Publishing Society of New Jersey, 1903.

Lender, Mark Edward. *One State in Arms: A Short Military History of New Jersey*. Trenton: New Jersey Historical Commission, 1991.

Lurie, Maxine, and Marc Mappen, eds. *Encyclopedia of New Jersey*. New Brunswick, NJ: Rutgers University Press, 2004.

Lurie, Maxine, and Richard Veit, eds. *New Jersey: A History of the Garden State*. New Brunswick, NJ: Rutgers University Press, 2012.

Mappan, Marc. *Jerseyana: The Underside of New Jersey History*. New Brunswick, NJ: Rutgers University Press, 1992.

Morris, Richard Knowles. *John P. Holland, 1841–1914. Inventor of the Modern Submarine*. Columbia: University of South Carolina Press, 1998.

Morrison, Samuel Eliot. *The Battle of the Atlantic: September 1939–May 1943*. Boston: Little, Brown, 1984.

Noble, Dennis L. *The Beach Patrol and Corsair Fleet: The US Coast Guard in World War II*. Washington, D.C.: Coast Guard Historian's Office, 1992.

Offley, Ed. *The Burning Shore: How Hitler's U-Boats Brought World War II to America*. Philadelphia: Basic Books, 2004.

Pike, Helen C. *Asbury Park's Glory Days: The Story of an American Resort*. New Brunswick, NJ: Rutgers University Press, 2005.

Powell, George R. *History of Camden County*. Philadelphia: L.J. Richards & Company, 1886.

Roberts, Russell, and Rich Youmans. *Down the Jersey Shore*. New Brunswick, NJ: Rutgers University Press, 1993.

Salter, Edwin. *A History of Monmouth and Ocean Counties*. Bayonne, NJ: E. Gardner & Sons, 1890.

Salter, Edwin, and George C. Beekman. *Old Times in Old Monmouth*. Freehold, NJ: Monmouth Democrat, 1874.

Simon, Bryant. *Boardwalk of Dreams: Atlantic City and the Fate of Urban America*. New York: Oxford University Press, 2004.

State of New Jersey. *Record of Officers and Men of New Jersey in Wars 1791–1815*. Trenton, NJ: State Gazette Publishing, 1909.

Taylor, Theodore. *Fire on the Beaches*. New York: W.W. Norton & Company, 1958.

Thomas, Lowell. *Raiders of the Deep*. Boston: Garden City, NY: Doubleday, Doran & Company, 1928.

Tuttle, Brad R. *How Newark Became Newark: The Rise, Fall and Rebirth of an American City*. New Brunswick: Rivergate Press, 2009.

United States Navy. *German Submarine Activities on the Atlantic Coast and Canada*. Washington, D.C.: U.S. Navy, 1920.

Werner, Herbert A. *Iron Coffins: A Personal Account of the German U-Boat Battles of World War II*. New York: Holt, Rinehart & Winston,1969.

Williamson, Gordon. *U-Boats of the Kaiser's Navy*. Oxford, UK: Osprey, 2002.

Woolf, Daniel. *4th of July, Asbury Park*. New York: Bloomsbury Publishing, 2005.

ARTICLES

Egan, M.F. "The Irish Inventor of the Modern Submarine." *Irish Monthly* 44, no. 513 (March 1916).

Gilbert, Troy. "The Coastal Picket Force." *BoatU.S.* (October/November 2015).

Moore, Samuel Taylor. "Britannia's Nightmare is an Irishman's Dream." *Knoxville (TN) News-Sentinel*, March 17, 1940.

Morris, Richard Knowles. "John P. Holland and the Fenians." *Journal of the Galway Archaeological and Historical Society* 31, no. 1/2 (1964/1965).

Ragan, Mark K. "A Union Whale Surfaces in New Jersey." *America's Civil War* (May 2008).

Strum, Harvey. "South Jersey and the War of 1812." *Cape May County Magazine of History and Genealogy* (1987).

Veit, Chuck. "The Innovative, Mysterious *Alligator*." *Naval History* (August 2010).

Watson, J. Francis, and William E. Watson. "Carl Krepper, American Pastor and Nazi Saboteur." *Lutheran Quarterly* 23 (2009).

Whitman, Edward C. "John Holland, Father of the Modern Submarine." *Undersea Warfare* (Summer 2003).

DOCUMENTS

Report, "CO, Co. F, 22nd Infantry, Subject: Submarine Attack, August 2, 1918." RG 156 entry 5958 box 1 file 052, NA. Copy courtesy of Cory J. Newman.

NEWSPAPERS

Altoona (PA) Tribune

Asbury Park Press

Boston Journal

Brooklyn Daily Eagle

Cincinnati Enquirer

Decatur (IL) Daily Review

Havre (MT) Daily News

Maysville (KY) Evening Bulletin

Nashua (NH) Telegraph

New York Herald

New York Times

Salt Lake (UT) Tribune

Scranton (PA) Republican

Wilkes-Barre (PA) Record

ONLINE SOURCES

Eastern Sea Frontier and U-Boat Records. http://www.uboatarchive.net/.

Monmouth County Oral History. http://www.visitmonmouth.com/oralhistory/Interview.htm.

Naval History and Heritage Command. http://www.history.navy.mil/.

New Jersey Governor's War Cabinet Notes. http://www.nj.gov/state/archives/szwaa001.html.

New Jersey Scuba Diving Wreck Sites. http://njscuba.net/sites/site_black_sunday.php.

NOOA Risk Assessment Package http://sanctuaries.noaa.gov/protect/ppw/pdfs/cayru.pdf

World War II Ships Register. http://www.warsailors.com/index.html.

INDEX

ABOUT THE AUTHORS

Joseph G. Bilby received his BA and MA degrees in history from Seton Hall University and served as a lieutenant in the First Infantry Division in Vietnam in 1966–67. He is assistant curator of the New Jersey National Guard and Militia Museum in Sea Girt, a member of and publications editor for the New Jersey Civil War Heritage Association, a columnist for the *Civil War News* and *New Jersey Sportsmen News* and a freelance writer, historian and historical consultant. He is the author, editor or coauthor of over four hundred articles and nineteen books on New Jersey, the Civil War and firearms history, including *Monmouth Courthouse: The Battle that Made the American Army*, coauthored with daughter Katherine, which was a Military Book Club selection. His most recent work includes editing the award-winning *New Jersey Goes to War* and coauthoring *350 Years of New Jersey History: From Stuyvesant to Sandy*, *Hidden History of New Jersey at War* and *On This Day in New Jersey History*. Bilby has received the Jane Clayton award for contributions to Monmouth County (NJ) history, an award of merit from the New Jersey Historical Commission for his contributions to the state's military history and the New Jersey Meritorious Service Medal from the state's Division of Military and Veterans Affairs.

Harry Ziegler was born in Neptune, New Jersey. He received his BA in English from Monmouth University and his MEd from Georgian Court University. He worked for many years for the *Asbury Park Press*, one of New Jersey's largest newspapers, rising from reporter to bureau chief to editor

and managing editor of the paper. He is currently associate principal of Bishop George Ahr High School in Edison, New Jersey, and is coauthor of *Asbury Park: A Brief History, Hidden History of New Jersey, Asbury Park Reborn, 350 Years of New Jersey History: From Stuyvesant to Sandy, Hidden History of New Jersey at War* and *On This Day in New Jersey History*.